FUNDAMENTALS

AND APPLICATION OF

Centrifugal Pumps

FOR THE PRACTICING ENGINEER

FUNDAMENTALS

AND APPLICATION OF

Centrifugal Pumps

FOR THE PRACTICING ENGINEER

BY

Alfred Benaroya

BSME, MSME, P.E.

Member of the American Society of
Mechanical Engineers (ASME)
Member of the Society of American
Military Engineers

PPC BOOKS

PETROLEUM PUBLISHING COMPANY
Book Division
Tulsa, Oklahoma
1978

PPC Books Division
The Petroleum Publishing Co.
1421 South Sheridan (P. O. Box 1260)
Tulsa, Oklahoma 74101, U.S.A.

Library of Congress Catalog Card Number: 77-78185
International Standard Book Number: 0-87814-040-9
Printed in U.S.A.

2 3 4 5-82 81 80

Preface

This book dwells exclusively on the fundamentals required for the efficient design of systems applying Centrifugal Pumps. It provides answers to most of those unanswered basic questions, and attempts to correct embedded misconceptions and misinterpretations, carried along for years, and reflected in inefficiently operating pumping systems.

The reader will not find descriptions, evaluations, construction details, cross-sections, and photographs of the numerous types of Centrifugal Pumps, since these are not deemed necessary for the scope of this book, and since they can be found in many excellent textbooks, handbooks, manufacturers' catalogs, and articles in technical magazines. On the other hand, he will be guided, step by step, through all the required fundamentals, without the thorough understanding of which a sound and efficient design of pumping systems employing Centrifugal Pumps would be inconceivable.

Examples, so desperately needed, and not readily available in the technical literature in this specific order, accompany each group of introduced and developed concepts and formulae; ample use is made of graphs, charts, tables, system head and head-capacity curves, in order to illustrate how to properly apply them. Appropriate comments accompany obtained results and solutions, and, when feasible, alternatives are investigated in detail.

Pumping system problems are resolved in their entirety, some of them taken from executed projects, all of them practical and applicable.

Finally, the introduction of changes into existing pumping systems employing centrifugal pumps, and the consequences of uncorrected deviations resulting from the introduced changes are investigated, discussed, and commented upon in detail. This represents a first in the literature, as far as the systematics of the presentation is concerned.

It is assumed that the reader has, to some extent, dealt with hydraulics, pumps, and pumping systems, and is, therefore, aware of the existence of the numerous charts, graphs, tables, nomograms, diagrams about friction losses, Reynold's Number versus friction factor, "K" factors, viscosities, velocities in pipes, piping data, steam tables etc., found in text and handbooks, manufacturers' brochures, articles in technical magazines and papers. None of this information, so abundantly available, is reproduced in this book, although ample use has been made of it.

A simple and easy to follow format of presentation is maintained. The unit dimensions of each parameter are repeated several times, whenever use is made of it in formulae. It is felt that the repetition of the dimensions helps the reader to reanalyze the formula constructively on the spot, without having necessarily to look up for its deduction elsewhere, and thus contributes substantially toward a smoother and more solid absorption of it on his part. Only numbers are used to refer to chapters, paragraphs, subparagraphs, figures, charts, tables, graphs, and examples.

Graphic and tabulated presentations are numbered consecutively, without differentiating, as far as their numbering is concerned, whether a figure follows a chart, or a table a graph. The chapter number and the sequential number of each of these are separated by a point. This system facilitates a rapid location of a referred-to figure, graph, or chart. (fig. 10.03, chart 10.04, table 10.05, etc., means that in chapter 10, figure 03 precedes chart 04, and table 05 follows chart 04).

Formulae carry the number of the chapter and their sequential number, separated by a dash. (2-07 means: the seventh formula in chapter 2). Examples, too, carry the number of the chapter and their sequential number, separated by a colon. (7:12 means: the twelfth example in chapter 7).

The chapter number is repeated on the top of each page in order to facilitate fast retrieval.

All formulae, with their specific reference number, are extracted and listed separately at the end.

Recommended literature is listed at the end, preceding the index.

The detailed index was prepared with the intention to provide the reader with instant reference to the specific page number, or numbers, never referring him to a different term in it. For example, "Discharge pressure" appears both under the letter D (Discharge pressure. . . . page XX), and the letter P (Pressure, discharge . . . . page XX).

Finally, I wish to extend my thanks to the following centrifugal pump manufacturing companies who granted the right to reproduce some of their pump head-capacity curves:

—Weinman Pump—LFE Corp., Fluid Control Div.
—Goulds Pump Inc., Seneca Falls, N.Y. 13148
—Pacific Pumping Co., Oakland, Ca.
—Aurora Pump—A Unit of General Signal Corp.
—Crane Deming Pumps
—Afton Pumps, Inc.
—Johnston Pump Company
—United Centrifugal Pumps

New York, 1978 The Author

Contents

CHARTS, TABLES AND ILLUSTRATIONS

Symbols

A	Cross-sectional area (square feet, or square inches)
Ans.	Answer
ATM	Atmospheric
BBL	Barrel (=42 Gallons)
BHP	Brake Horse Power
BPH	Barrels per Hour
CF	Cubic Feet
CFM	Cubic Feet per minute
CFS	Cubic Feet per second
CP	Centipoise, unit to measure viscosity
CS	Centistoke, unit to measure viscosity
CU FT	Cubic Feet
D, d	Pipe diameter
D	Impeller Diameter
dia	Diameter
E	Energy
e	Absolute pipe roughness, in feet
E_p	Pump Energy
eff	efficiency
f	friction factor, dimensionless
°F	temperature degree, Fahrenheit
FPS	feet per second
FT.	feet, foot
FT-LBS	foot-pounds
G	Gallon
g	gravitational constant, 32.2 ft/sec^2
GPM	Gallons per minute
H	Total Head, in feet
H_f	Head loss due to friction, in feet
HP	Horsepower
i	Interest Rate, in percent (fraction)
ID	internal pipe diameter

In.	inch(es)
K	resistance coefficient for valves, fittings, etc., dimensionless
L	pipe length, in feet
L	salvage value of equipment after n years
Le, le	equivalent length, in feet.
lb(s)	pound(s)
Max	maximum
Min	Minimum
N	Rotation speed (RPM)
n	number of equivalent pipes to substitute one pipe
n	life of an asset, in years
NPSH	Net Positive Suction Head
NPSHA	Available Net Positive Suction Head
NPSHR	Required Net Positive Suction Head
NS	Specific Speed
OC	Annual operating cost of an installation, in $
OD	Outside diameter of a pipe
P	Pressure, in feet, PSIA or PSIG
P	Initial cost of an installation, in $
PR	Pressure
PSI	pounds per square inch
PSIA	pounds per square inch absolute
PSIG	pounds per square inch gauge
Q	Capacity, Flowrate, in GPM or BPH for liquids
RE	Reynold's Number
RPM	Revolutions per minute
Sp.gr.	Specific gravity
Sq ft.	Square feet (foot)
Sq in.	Square inch(es)
SSU	Saybolt Seconds Universal (unit to measure viscosity)
T	Temperature, Torque
V	Velocity, normally flow velocity through a pipe, in FPS
V_h	Velocity head, in FT
V_p	Vapor Pressure, in PSIA
Z	Static elevation, in FT

1.

The Parameters

1.1 PRESSURE

The following units are used in the industry to designate pressure developed by a liquid:

1.1.1. Pounds per square inch, absolute (PSIA)
1.1.2. Pounds per square inch, gauge (PSIG)
1.1.3. Pounds per square inch (PSI)
1.1.4. Head, or Column of liquid (FEET, or INCHES)
1.1.5. Vacuum. Although not a unit, it is accepted as one, and used to indicate pressure below atmospheric.

Some confusion still reigns when using any of above units, especially when they are quoted loosely, like "pounds," without specifying PSIA, PSIG, or "feet," without specifying the liquid, "inches vacuum," without specifying mercury, or water, and the atmospheric pressure.

This superficiality may at times create misunderstandings, cause errors, resulting in costly misinterpretations. It is, therefore, mandatory to understand the meaning of each unit, and learn to be explicit when using it.

1.1.1 PSIA—This is measured from a pressure level equivalent to ABSOLUTE ZERO. Thus, at sea level, atmospheric pressure amounts to 14.7 PSIA. At elevation 4,000 FT it amounts to 12.7 PSIA, and at 10,000 FT to 10.1 PSIA.

The literature provides formulae and tables correlating elevations above sea level and atmospheric pressure in PSIA. In many applications sea level atmospheric pressure is used.

1.1.2 PSIG—A pressure gauge indicates pressure above atmospheric. Therefore, it will indicate ZERO if atmospheric is all the pressure existing, independent of the elevation above sea level. In this particular case PSIA is equivalent to atmospheric pressure. Thus, the following relationship has been established:

PSIA = Atmospheric Pressure + PSIG 1–01

1.1.3 PSI—This indicates pressure in general, when no reference is made to absolute or gauge pressure. The difference between two gauge readings is expressed in PSI, and not, as often encountered, in PSIG. (200 PSIG − 120 PSIG = 80 PSI). The pressure loss due to friction in a pipe, or in equipment, is expressed in psi per unit length, or respectively in psi for the unit for a given flowrate.

1.1.4 Head, or Column of Liquid—Heads, or heights of columns of liquid, are generally expressed in FEET or INCHES. The pressure developed by the liquid on the base of the column depends on the height of the column and the specific gravity of the liquid. Water at 68° F has a specific gravity of 1.0. A column 2.31 FT high will develop a pressure of 144 lbs/sq ft = 1 PSI. (The density of water at this temperature is equivalent to 62.33 lbs/CU FT; then 2.31 FT × 62.33 lbs/CU FT = 144 lbs/sq ft = 1 lb/sq in).

1 PSI = 2.31 FT of water at 68°F 1–02
0.433 PSI = 1.00 FT of water at 68°F 1–03
1 PSI (any liquid) = 2.31/sp. gr. of liquid (FT) 1–04
[1 Foot (any liquid) = sp. gr. of liquid/2.31 (PSI)]

- EXAMPLE 1:1

Liquid	*Sp. Gr.*	*1 PSI equals:*
Gasoline	0.55	4.2 FT of gasoline
Brine	1.20	1.925 FT of brine
Mercury	13.60	0.17 FT (= 2.04 in.) of mercury
Water	1.00	2.31 FT (= 27.72 in.) of water

- EXAMPLE 1:2

Atmospheric pressure at sea level (14.7 PSIA) in FT of liquid:

Water	14.7 × 2.31	= 33.96 FT (407.52 in.) of water
Gasoline	" × 4.2	= 61.74 FT of gasoline
Brine	" × 1.925	= 28.30 FT of brine
Mercury	" × 0.17	= 2.50 FT (30 in.) of mercury

From examples 1:1 and 1:2 we obtain:

1 in. of water = 0.036 PSI 1–05
1 in. of mercury = 0.490 PSI 1–06

1.1.5 Vacuum—This indicates ZERO absolute pressure, i.e. ABSOLUTE EMPTINESS.

VACUUM = 0 PSIA 1–07

If, as usually happens, we only have partial vacuum, one definitely needs to know the atmospheric pressure, in order to express partial vacuum meaningfully.

Partial vacuum is measured from atmospheric pressure down. At atmospheric pressure we define vacuum to be equal to ZERO percent (0%), and at 0 PSIA to 100%.

The following simple chart illustrates the relationships defined above:

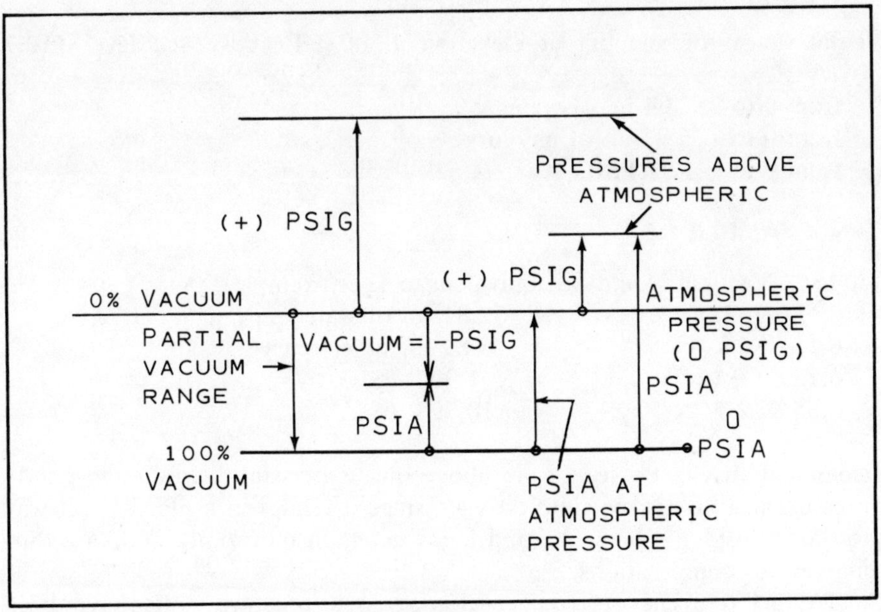

CHART 1.01

Diagrammatic representation of absolute, gauge pressure and vacuum.

Units to measure partial vacuum:

Inches of Water
Inches of Mercury
Negative Gauge (− PSIG)

■ EXAMPLE 1:3

Total range of vacuum at sea level (14.7 PSIA):

from 0 to 407.52 in. of water
from 0 to 30 in. of mercury
from 0 to −14.7 PSIG

■ EXAMPLE 1:4

A 15% vacuum would be equivalent to (per example 1:3):

0.15×407.52	=	61.13 in. of water
0.15×30	=	4.5 in. of mercury
0.15×(−14.7)	=	− 2.205 PSIG
0.85×14.7	=	+12.495 PSIA; (85% of PSIA at atm. pressure)

■ EXAMPLE 1:5

Total range for vacuum at elevation 4,000 FT above sea level (12.7 PSIA):

from 0 to 352.04 in. of water
from 0 to 25.91 in. of mercury
from 0 to −12.7 PSIG

■ EXAMPLE 1:6

A 15% Vacuum would be equivalent to (per example 1:5):

0.15×352.04	=	52.81 in. of water
0.15× 25.91	=	3.89 in. of mercury
0.15×(−12.7)	=	− 1.905 PSIG
0.85×12.7	=	+10.795 PSIA

Comment: It can be seen from above that expressing below atmospheric pressure in PSIA is the simplest way, since it relates in a more direct way to the atmospheric pressure, and helps maintain uniformity and ease during various computations.

To facilitate easy reference, all discussed pressure units have been tabulated in TABLE 1.02.

1.2 VOLUME

Basic Units:

1 Cubic FT (CF)	=	7.481 Gallons	1–08
1 Cubic FT	=	0.178 Barrel	1–09
1 Gallon (G)	=	0.1337 CF	1–10
1 Barrel (BBL)	=	5.615 CF	1–11
1 BBL	=	42 Gallons	1–12
1 Gallon	=	0.0238 BBL	1–13

1.3 FLOWRATE (Q = VOLUME PER UNIT TIME)

Basic Units:

CF per second (CFS)
CF per minute (CFM)

TABLE 1.02. COMPARATIVE TABLE OF PRESSURE UNITS AND READINGS

	psi	psia	psig	ft. of water	in. of water	in. of mercury	% vacuum	
EQUIVALENCE:								
1 psi	1.00	/	/	2.31	27.72	2.04	/	
1 ft. of water	0.433	/	/	1.00	12.00	0.883	/	
1 in. of water	0.036	/	/	0.0833	1.00	0.0736	/	
1 in. of mercury	0.490	/	/	1.132	13.60	1.00	/	
READING:								
0 psia	/	0	−14.7	33.96 VACUUM	407.52 VACUUM	30 VACUUM	100	ATM. PR. =14.7 psia
0 psia	/	0	−12.0	27.72 VACUUM	332.64 VACUUM	24.46 VACUUM	100	ATM. PR. =12 psia
0.5 psia	/	0.5	−14.2	32.80 VACUUM	393.62 VACUUM	28.94 VACUUM	96.6	ATM. PR. =14.7 psia
1.0 psia	/	1.0	−13.7	31.65 VACUUM	379.80 VACUUM	27.93 VACUUM	93.2	ATM. PR. =14.7 psia
14.5 psia	/	14.5	− 0.2	0.46 VACUUM	5.52 VACUUM	0.41 VACUUM	1.4	ATM. PR. =14.7 psia
14.5 psia	/	14.5	+ 2.5	5.77 GAUGE	69.24 GAUGE	5.1 GAUGE	/	ATM. PR. =12.0 psia
14.7 psia	/	14.7	0	33.96 ABSOLUTE	407.52 ABSOLUTE	30 ABSOLUTE	0	ATM. PR. =14.7 psia
15.7 psia	/	15.7	+ 1.0	2.31 GAUGE	27.72 GAUGE	2.04 GAUGE	/	ATM. PR. =14.7 psia
5% vacuum	/	13.965	− 0.735	1.70 VACUUM	20.4 VACUUM	1.50 VACUUM	5	ATM. PR. =14.7 psia
3″ vacuum-mercury	/	13.230	− 1.47	3.4 VACUUM	40.80 VACUUM	3.00 VACUUM	10	ATM. PR. =14.7 psia
10″ vacuum-water	/	14.34	− 0.36	0.833 VACUUM	10.00 VACUUM	0.73 VACUUM	2.45	ATM. PR. =14.7 psia
5 psig	/	16.60	5.00	11.55 GAUGE	138.6 GAUGE	10.19 GAUGE	/	ATM. PR. 11.6 psia

G per minute (GPM)
BBL per hour (BPH)

Note: Although imprecise, the literature uses "Capacity" instead of the correct term "Flowrate".

The following Table 1.03 correlates flowrate units:

TABLE 1.03. CORRELATION OF FLOWRATE UNITS

	CFS	CFM	GPM	BPH
1 CFS	1.00000	60.0000	448.860	640.80
1 CFM	0.01670	1.0000	7.481	10.68
1 GPM	0.00223	0.1337	1.000	1.43
1 BPH	0.00156	0.0936	0.700	1.00

1.4 FLOWRATE (W = WEIGHT PER UNIT TIME)

Basic Unit: Pounds per hour (LBS/HR)

1.5 DENSITY (WEIGHT PER UNIT VOLUME)

Basic Unit: Pounds per Cubic Foot (LBS/CF)

■ EXAMPLE 1:7

Density of Water at 68° = 62.33 LBS/CF 1–14

1.6 SPECIFIC GRAVITY (SP.GR.)

Ratio of the density of a fluid to that of water.

■ EXAMPLE 1:8

Sp. Gr. of water at 68°F = 62.33/62.33 = 1.00 1–15

1.7 VELOCITY (DISTANCE PER UNIT TIME)

Basic Unit: Feet Per Second (FPS)

1.8 HORSEPOWER (HP = WORK PER UNIT TIME)

1 HP = 550 ft-lbs per sec = 33,000 ft-lbs per min 1–16

1.9 TORQUE (T = MECHANICAL MOMENT CAUSING ROTATION)

Basic Unit: pound-feet (LB-FT)
TORQUE = (T) = 5,250 × HP/RPM 1–17
Where RPM = revolutions per minute.

1.10 VISCOSITY OF LIQUIDS

Viscosity indicates the degree of internal friction of a liquid. The higher the viscosity of a specific liquid, the higher the resistance encountered during its flow through a pipe, valve, pump, etc.

Units used are Saybolt Seconds Universal (SSU), Centipoise, absolute (CP), Centistoke, kinematic (CS)

1.10.1 SSU

This represents the number of seconds required for 60 cubic centimeters to flow out through a short tube at a given temperature. The more viscous the liquid, the higher the SSU number. Water at 60°F = 31.5 SSU, Oils = 50 to 4,000 SSU.

1.10.2 CP

Metric units are used to express CP (Dyne-sec/cm^2). It will suffice to state that the viscosity of water at 68°F = 1 CP. A liquid which has a viscosity of 120 CP is 120 times more viscous than water.

1.10.3 CS

This viscosity unit is called kinematic, because force units are not involved; only length and time units (FT^2/SEC). The relationship between CS and CP:

Centistoke = Centipoise/Specific Gravity 1–18

Water at 68° F = 1 CS = 1 CP; (sp. gr. of water at 68° F = 1).

Note: The viscosity of a liquid is inversely proportional to its temperature: the higher the temperature, the lower the viscosity, and vice versa.

1.11 HEAD

(Refer to item 1.1.4)

The height of a liquid above or below a fixed level is defined as STATIC head, and is measured in feet. It is positive when liquid level is above, and negative when it is below this fixed level.

■ EXAMPLE 1:9

A storage tank has a liquid level 30 feet above a nozzle in the same tank. The static head at the nozzle = + 30 feet.

The centerline of a pump is 20 feet above the liquid level of a tank. The static head at the centerline of the pump: minus 20 feet.

The literature, in order to avoid negative head, has introduced the term STATIC LIFT, which equals negative static head.

Note: In this book we will not use the term LIFT. We will only use positive or negative static head. This makes computations easier and eliminates mistakes.

1.12 VELOCITY IN A PIPE, V (IN FPS)

Refer to item 1.7. The velocity of a liquid in a pipe is a function of the flowrate (= capacity) Q, and the cross-section of the pipe (A).

$V = Q/A = 4Q/(\pi D^2) = 1.273\ Q/D^2$ 1–19
(V in FPS; Q in CFS; D in FT)

In conjunction with TABLE 1.03:

$V = 1.273\ Q/(448.86\ D^2) = Q/(352.6\ D^2)$ 1–20
(V in FPS; Q in GPM; D in FT)

$V = Q \times 144/(352.6 \times D^2) = Q/(2.448 \times D^2)$ 1–21
(V in FPS; Q in GPM; D in inches)

$Q = 2.448 \times D^2 \times V$ (V in FPS; Q in GPM; D in inches) 1–22

▪ EXAMPLE 1:10

Compute the velocity of a liquid flowing in an 8-in. pipe with an internal diameter of 7.981 inches (= 0.665 FT), and a cross-sectional area of 50 square inches (= 0.3472 sq ft). The flowrate = 1,200 GPM (2.673 CFS).
Solution: with 1–19:

$\mathbf{V = Q/A}$ = CFS/SQ FT = 2.673/0.3472 = 7.70 FPS

also with 1–19:

$$\mathbf{V = 1.273\ Q/D^2} = \text{CFS/FT}^2 = \frac{1.273 \times 2.673}{0.665^2} = 7.70 \text{ FPS}$$

with 1–20:

$\mathbf{V = Q/(352.6\ D^2)}$; in FPS; $= 1{,}200/(352.6 \times 0.665^2) = 7.70$ FPS

with 1–21:

$\mathbf{V = Q/(2.448 \times D^2)}$; in FPS; $= 1{,}200/(2.448 \times 7.981^2) = 7.70$ FPS

1.13 VELOCITY HEAD

The kinetic energy in a liquid at any point is defined as velocity head, and is measured in feet of liquid.

Velocity head $= V_h = V^2/2g$ 1–23

V = velocity in FPS

g = acceleration due to gravity = 32.174 FT/sec²

The velocity head could also be considered equivalent to the distance the liquid has to fall to or from, in order to attain the velocity V.

■ EXAMPLE 1:11

The flow velocity in a pipeline is 10 FPS. Compute the velocity head.
$V_h = V^2/2g = 10^2/(2 \times 32.2) = 1.55$ FT Ans.

■ EXAMPLE 1:12

Compute the flowrate through pipes with different internal diameters for all of which the velocity head is 1.55 FPS (V = 10 FPS). Use formulae 1–22 and 1–23, and tabulate the results.

OD (in.)	ID (in.)	V (FPS)	$V^2/2g$ (FT)	Q (GPM)
10.750	10.02	10	1.55	2,458
12.750	12.00	10	1.55	3,525
16.000	15.25	10	1.55	5,693
20.000	19.18	10	1.55	9,005

Note: The Velocity head, as can be seen from the formula, depends exclusively on the flow velocity which, in turn, is a function of the pipe diameter and capacity.

■ EXAMPLE 1:13

Compute the velocity heads for various capacities through the same 10 inch pipe (ID = 10.02 in.)

Q (GPM)	V (FPS)	$V^2/2g$ (FT)
500	2.04	0.065
1000	4.07	0.26
1500	6.10	0.58
2000	8.14	1.03
3000	12.20	2.30

The velocity head for the same size pipeline increases with the capacity and the velocity.

1.14 VAPOR PRESSURE.

The vapor pressure of a liquid at a given temperature is that pressure at which the liquid starts boiling and vaporizing. The boiling point and the respective vapor pressure of a specific liquid are a function of its temperature. It follows that a liquid has different vapor pressures and boiling points for different temperatures.

"Boiling Point" doesn't necessarily mean that the liquid is very hot. This, though, is true in certain cases, as indicated in the following examples.

■ EXAMPLE 1:14

Consider a liquid in a closed container.
P = pressure in the container, in PSIA
T = temperature of the liquid, in °F
Vp = vapor pressure of the liquid, in PSIA
Liquid = water.

	P	T	Vp (FROM STEAM TABLES)
1.14.1	14.7	212	14.7
1.14.2	5.0	162	5.0
1.14.3	150.0	358	150.0
1.14.4	0.1	34	0.1

In 1.14.1, the pressure is atmospheric. When heated, the water will start boiling at 212° F. By definition then, the vapor pressure of water at 212° F is 14.7 PSIA. In other words, in order for the water to boil at 212° F, the pressure will have to be 14.7 PSIA. The water will not boil for temperatures less than 212° F, while the pressure is 14.7 PSIA.

In example 1.14.2, partial vacuum exists in the container, i.e. the pressure is below atmospheric. When heated, the water will start boiling at 162° F. Therefore, the vapor pressure of water at 162° F is 5 PSIA.

In example 1.14.3, the container is pressurized, i.e. the pressure is above atmospheric. This means that, if the pressure outside the vessel is 14.7 PSIA (atmospheric at sea level), a gauge will indicate 135.3 PSIG in the vessel (=150–14.7). When heated, the water will start boiling only when the temperature will reach 358° F. Therefore the vapor pressure for water at 358° F is 150 PSIA.

It follows that, depending on the pressure in the container, the water will start boiling at a lower or higher temperature. As a matter of fact, if the pressure in the container is very low, water will start boiling at a temperature which one could consider freezing. Example 1.14.4 indicates that for a pressure of 0.1 PSIA water will start boiling at 34° F which is only 2 degrees above freezing under atmospheric conditions. The vapor pressure for water at 34° F is very low: 0.1 PSIA.

Consider now a container with water at 60 degrees F in it. The pressure in it is 14.7 PSIA (atmospheric at sea level). The water is not boiling, because the temperature is below 212° F, per 1.14.1. What is, then, the vapor pressure for water at 60° F and atmospheric pressure of 14.7 PSIA?

Per above definitions, it would be the pressure at which water at 60° F will start boiling. Tables indicate this pressure to be 0.25 PSIA (= 1.7% of 14.7).

In case the pressure in the vessel is 1000 PSIA, and the temperature of the water is again 60° F, what would be the vapor pressure?

The boiling temperature for water pressurized at 1000 PSIA is, per tables, 544° F. Therefore the water at 60° is relatively cool. As above, the vapor pressure is again 0.25 PSIA, only this time it represents just 0.025% of the total pressure.

1.14.5. The pressure in a closed container is 20 PSIA (= 5.3 PSIG for an atmospheric pressure of 14.7 PSIA). The water temperature is 200° F. Tables indicate that, in order to boil, the water temperature must reach 228° F. Thus, at 200° F, the water is not yet boiling, and it is not vaporizing.

On the other hand, in order to boil at 200° F, the pressure in the container must be, per tables, 11.5 PSIA. If we wish to prevent 200° F water from boiling and vaporizing, we have to maintain the pressure at a level higher than 11.5 PSIA.

1.14.6. The pressure in a boiler is 1800 PSIA. The temperature is 550° F. Boiling and vaporization will start at 621° F. From tables we know that the vapor pressure of water at 550° F is 1050 PSIA, a substantial part of the total pressure (58.3%) In order for the 550° F water to boil, one either has to lower the pressure from 1800 PSIA to 1050 PSIA, or increase its temperature from 550° to 621° F.

1.14.7. A sudden lowering of the pressure in a container below the vapor pressure of the liquid in it at the specific temperature will cause a portion of the liquid to vaporize instantaneously till equilibrium is reached. It is said that flashing will take place.

If in example 1.14.5 the pressure suddenly drops from 20 PSIA to 10 PSIA without appreciable change in temperature (200° F), the water will boil off, and a portion of it will vaporize, because the boiling temperature for water at 10 PSIA is 193° F, 7 degrees lower than 200° F. Part of the container will fill up with steam. The amount of vaporization (flashing) is proportional to the differential between the actual and boiling temperatures.

In the case of a centrifugal pump, taking suction from this container under these circumstances, vapors (steam) will fill a portion of the suction pipe, a condition considered detrimental to the performance and life of the pumps. It is said that the pump "cavitates." Therefore precautions have to be taken to avoid these conditions. Tables indicate that various liquids have different vapor pressures for the same temperature:

Temperature: 70° F	Vapor pressure (PSIA)
Water	0.36
Kerosene	0.49
Methylalcohol	2.00
Ethyl Chloride	4.40
Methyl Chloride	80.00
Carbon Dioxyde	900.

The literature, as well as product catalogs, provide information about vapor pressures at different temperatures, vital information for the proper design of a pumping system.

1.15 REYNOLD'S NUMBER (RE)

The Reynold's Number is DIMENSIONLESS. It is an indicator of the flow pattern for pipes flowing full. It is a function of the internal diameter, flowrate, and the viscosity of the liquid. There are several available formulae in the literature for RE. With respect to the scope of this book, we will use only one:

RE = 3162 × Q/(D × CS) 1–24

Q = flowrate (capacity), in GPM

D = internal pipe diameter, in INCHES

Cs = Viscosity, in Centistokes (see 1.10.3)

A flow pattern is defined as turbulent when the value of RE exceeds 2,100. A flow pattern is defined as laminar when the value of RE is less than 2,100.

It is accepted to characterize the zone 2,100 to 4,000 as undefined, and some scientists prefer to consider the range as laminar.

In the following example, Reynold's Numbers have been computed for various conditions and tabulated for easy review.

■ EXAMPLE 1:15

FLUID	WATER	SULFURIC ACID	CRUDE OIL	ACETONE	FUEL OIL
Sp. gr.	1.0	1.50	0.85	0.79	0.90
CS	1.0	4.66	5	0.50	400
Q (GPM)	250	100	59,000	600	3,000
Nom. Dia	4″	3″	48″	6″	12″
ID	4.026″	3.068″	46″	6.065″	12″
RE	196,349	22,116	811,122	625,622	1,976
FLOW PATTERN	TURB.	TURB.	TURB.	TURB.	LAM.

2.

Pressure Loss Due to Friction

Pressure is lost due to friction, when a liquid:

2.1. Exits a vessel, or storage tank into a pipe.

2.2. Flows through valves, strainers, and other equipment.

2.3. Flows through pipe bends and other piping fittings.

2.4. Flows through a gradual or sudden pipe enlargement, or contraction.

2.5. Enters a vessel, or storage tank from a pipe.

2.6. Flows through a straight pipe.

Various short terms are being used for pressure loss due to friction. They include "Friction loss", "Pressure Drop", "Friction Drop", and "Friction Head Loss". None of these are precise, but Friction loss is widely accepted and used.

Since head loss due to friction represents loss of energy and since energy of a flowing liquid is represented by its velocity head (item 1.13), it is accepted to express the loss as a multiple or part of it, by means of a resistance factor "K":

Friction loss = Energy Loss = $K \times V^2/2g$ (FT) 2–01

■ EXAMPLE 2:1

Refer to items 2.1 through 2.6 above:

2.1. Exit loss: $K = 0.1–1.00$.

2.2. Loss through valves: $K = 0.2–3.00$ (general range).

2.3. 90° bend: $K = 0.5–0.9$.

2.4. Enlargement, or contraction in pipes:

In this case, due to the two different pipe diameters, the energy will be represented by two different velocity heads. It can be proved that:

2.4.1. Loss due to sudden enlargement equals:

$K (V_1 - V_2)^2/2g$, where $K = 1.0$.

2.4.2. Loss due to gradual enlargement equals:

$K(V_1 - V_2)^2/2g$, where $K = 0.02$ to 0.8, depending on the angle of the enlargement.

2.4.3. Loss due to sudden contraction:

$K \times V^2/2g$, where V represents the higher velocity and $K = 0.1$ to 0.5, depending on the ratio of diameters.

2.5. Entrance Loss: $K = 1.0$.

2.6. Resistance factor "K" for pipes:

$K = f \times L/D$ 2–02

where f = friction factor, depending on the Reynold's Number (1–24) and the relative pipe roughness expressed as e/D. Friction factor charts, abundant in the literature, relate f, RE, and e/D. "f" values range from 0.009–0.08 and are dimensionless.

e = absolute roughness, in FT, represents the deviation from the theoretical internal pipe diameter in the form of internal projection. The rougher the pipe, the larger is "e". Values for "e" are:

Commercial steel, 0.00015 feet.

Cast iron, 0.00085 feet.

Concrete, 0.001 to 0.010 feet.

e/D is equivalent to the relative roughness, a dimensionless ratio between "e" (in feet) and pipe diameter D (in feet). For commercial steel, for a 12-inch pipe, e/D = 0.00015, and for a 36-inch pipe, e/D = 0.00003. In the preceding, L = length of pipe, in feet, and D = internal pipe diameter, in feet.

■ EXAMPLE 2:2

Refer to figure 2.01: a 12-inch diameter steel pipeline, 1,000 feet long, connects two storage tanks at different elevations. With two 12-in. gate valves, and ten 90° bends in the line, compute the friction loss for a flow velocity of 10 FPS, and a friction factor of $f = 0.02$; where $V^2/2g = 1.553$ FT.

Solution:

Exit Loss	$(1 \times 1) \times V^2/2g$	$= 1 \times V^2/2g$	$= 1.553$ FT
Two gate valves	$(2 \times 0.3) \times V^2/2g$	$= 0.6 \times V^2/2g$	$= 0.932$ FT
Ten 90° bends	$(10 \times 0.9) \times V^2/2g$	$= 9 \times V^2/2g$	$= 13.977$ FT
Entrance	$(1 \times 1) \times V^2/2g$	$= 1 \times V^2/2g$	$= 1.553$ FT
Pipe Friction Loss	$(0.02 \times 1000/1) \times V^2/2g$	$= 20 \times V^2/2g$	$= 31.060$ FT
Total	$= 11.6\ V^2/2g + 20\ V^2/2g$	$= 31.6\ V^2/2g$	$= 49.075$ FT Ans.

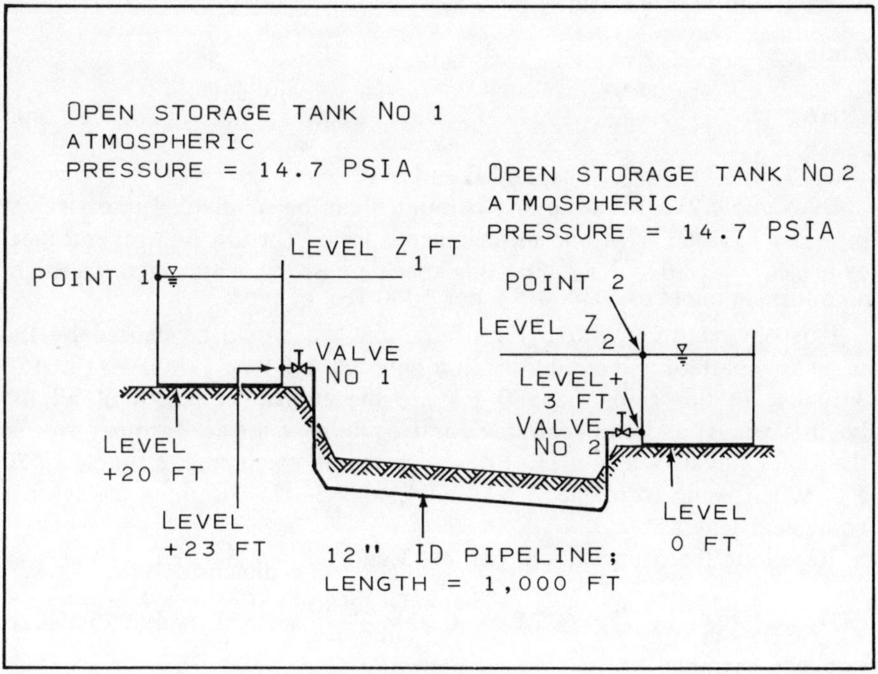

FIG. 2.01

2.6.1 FRICTION LOSS IN PIPES

The general formula for friction loss in pipes, in conjunction with formulae 2–01 and 2–02, results in the expression:

$$H_f = f \times \frac{L}{D} \times V^2/2g \text{ (FT)} \qquad \text{2–03}$$

This formula can be transformed as follows:

$$V = \frac{4Q}{\pi D^2} \qquad \text{(from 1–19)}$$

$$V = Q/(352.6 \times D^2) \qquad \text{(from 1–20)}$$

$V^2 = Q^2/(124{,}327 \times D^4)$, where V is in FPS; Q is in GPM; and D is in FT.

$V^2/2g = Q^2/(8{,}006{,}658 \times D^4)$, in FT.

In conjunction with 2–03:

$$Hf = f \times \frac{L}{D^5} \times Q^2/8{,}006{,}658 \qquad \text{2–04}$$

where L and D in FT and Q in GPM.

And:

$$Hf = f\times\frac{L}{D^5}\times\frac{Q^2}{32.18}\ldots\ldots 2\text{–}05$$

where L in FT; D in inches; and Q in GPM.

In example 2:2, the total loss computed can be subdivided into straight pipe loss (31.060 FT), and miscellaneous losses (bends, fittings, entrance, exit, etc., = 18.015 FT). For this specific case, we can say the straight pipe loss amounts to 31.060 FT per 1000 feet of pipe.

It is obvious that 18.015 FT of friction loss would be caused by the same flow pattern and conditions in a pipe 580 FT long (1000 × 18.015/31.060). In this case, the 580 FT are the equivalent length of all the bends; fittings; and exit, entrance, and equipment losses. In other words, the total equivalent length (Le) of the system in example 2:2 equals 1,580 FT. When using formulae 2–03, 2–04, and 2–05, one must always use equivalent lengths.

Repeating the computation, per 2–03 and Le = 1,580 FT:

$H_f = f\times\frac{L}{D}\times V^2/2g$ (FT) $= 0.02\times\frac{1{,}580}{1\ FT}\times 1.553 - 49.075$ FT, as in above example.

The literature provides tables with equivalent pipe lengths for equipment and fittings as a function of the pipe size. From the results obtained in the above example, one can figure out the equivalent lengths as follows:

Exit loss = 1.553 FT; 1.553×1,000/31.060 = 50 equivalent FT; (2) 12-inch gate valves = 0.932 FT; 0.932×1,000/31.060 = 30 equivalent FT. (10) 12-inch 90° bends = 13.977 FT; 13.977×1,000/31.060 = 450 equivalent FT. The entrance loss = Exit loss = 50 equivalent FT.

(Total = 50+30+450+50 = 580 FT)

Comparing some of these values with tabulated ones in the technical literature, we may find lower equivalent lengths than the ones computed above since we used conservative values for the factor K in example 2:1.

When computing equivalent lengths, the designer will have to use his judgement for each specific case. If pipe friction loss is the largest, one can afford to be conservative with respect to the fitting losses because they will not affect the total loss substantially. On the other hand, when the pipe-lengths are short and the number of fittings, valves, and other equipment items is large, extreme care has to be taken to obtain the specific K factors, or equivalent lengths. In cases of this kind, the equipment manufacturer must be consulted.

3.

Energy Loss

3.1. Conforming to the Principle of Conservation of Energy, Bernoulli's Equation for steady flow of incompressible liquids states:

$E_1 + E_A - E_L - E_E = E_2$

where:

E_1 = Energy at point 1 in a system.
E_A = Energy added, if any.
E_L = Energy lost.
E_E = Energy extracted, if any.
E_2 = Energy at point 2 in the same system.

The energy of an incompressible liquid, while flowing, consists of:

P = Pressure,
$V^2/2g$ = Kinetic Energy,
Z = relative position (elevation),
E_p = added energy (by a pump, for example),
H_f = lost energy (friction losses, etc.).

It is accepted to express all above parameters in feet. Thus:

$$P_1 + V_1^2/2g + Z_1 + E_p - H_f = P_2 + V_2^2/2g + Z_2 \qquad (3\text{–}01)$$

where indices 1 and 2 refer to locations in the system. In case the dimension units are not expressed in feet, they must be converted into feet before any computations are initiated.

■ EXAMPLE 3:1

Refer to figure 2.01. With valves Nos. 1 and 2 open, compute the flowrate through the 12-in. pipeline (fluid: water, sp. gr. = 1.00).

Solution: Reference Points: Point 1 and Point 2. With equation 3–01:

$P_1 + V_1^2/2g + Z_1 + E_p - H_f = P_2 + V_2^2/2g + Z_2$
$P_1 = 14.7\ \text{PSIA} = 14.7 \times 2.31 = 33.96\ \text{FT}$

V_1 = 0 (level at point 1 constant)
Z_1 = 50 FT
E_p = 0 (no energy added)
H_f = 31.6 $V^2/2g$ (from example 2:2)
P_2 = P_1 = 33.96 FT
V_2 = V_1 = 0
Z_2 = 30 FT
V = flow velocity in the 12-in. pipe

$33.96+0+50+0 - 31.6\ V^2/2g = 33.96+0+30$
$31.6\ V^2/2g = 20$
$V^2/2g = 0.633$ FT
$V = 6.38$ FPS

In conjunction with formula 1–22:

$V = Q/(2.448 \times D^2)$, FPS; (Q = GPM; D = in.) Q = 2,250 GPM Ans.

Comment: In example 2:2, the friction factor was assumed to be equal to 0.02. One could now refine its value by computing the Reynold's Number and read off the corresponding f value from charts in the literature.

$RE = 3{,}162\ Q/(D \times CS)$; (Q = GPM; D = in.) (1–24)
$RE = 593{,}000$

From charts, for a smooth, new steel pipe,
$(e/D = 0.0001)$, $f = 0.0132$.

In example 2:2, we had for the pipe friction loss:

$$0.02 \times \frac{1000}{1} \times V^2/2g = 20\ V^2/2g.$$

In this case, we will have:

$$0.0132 \times \frac{1000}{1} \times V^2/2g = 13.2\ V^2/2g$$

Total = 11.6 $V^2/2g$ (fitting loss per 2:2) + 13.2 $V^2/2g$ = 24.8 $V^2/2g$
Hf = 24.8 $V^2/2g$ = 20 (= $Z_1 - Z_2$ = 50 − 30)
$V^2/2g = 0.806$
$V = 7.21$ FPS
Q = 2,540 GPM (from 1–22) Ans.

Pipe Loss = 13.2×0.806 = 10.65 FT
Fitting Loss = 11.6×0.806 = 9.35 FT

Entrance Loss = $1 \times V^2/2g$	= 0.806
Two 12″ Gates = $0.6 \times V^2/2g$	= 0.484
Nine 12″ Bends = $9 \times V^2/2g$	= 7.254
Exit Loss = $1 \times V^2/2g$	= 0.806
Total Fittings Loss	= 9.35 (as above)

Equivalent length of the system in this case:

$$H_f = f\frac{Le}{D}\times V^2/2g = 20 = 0.0132\frac{Le}{1}\times 0.806$$

Le = 1,879 FT, as compared to 1,580 in the previous example. This is due to the difference in the friction factor "f" and the velocity head.

3.2. The factor f can be calculated for new, smooth pipes, in case the Reynold's number doesn't exceed 1,000,000, by means of the formula:

$$f = \frac{0.184}{RE^{0.2}} \quad \text{........................} 3\text{–}02$$

Using this formula for our case above (RE = 593,000):

$$f = \frac{0.184}{593{,}000^{0.2}} = 0.01289 \text{ (as compared to 0.0132)}$$

■ EXAMPLE 3:2 (Refer to fig 3.01)

For the conditions shown compute: The flowrate with both valves open and liquid levels in tanks constant; with only valve No. 2 closed, the static pressure at point B; and the energies at both points B and B′ during full flow:

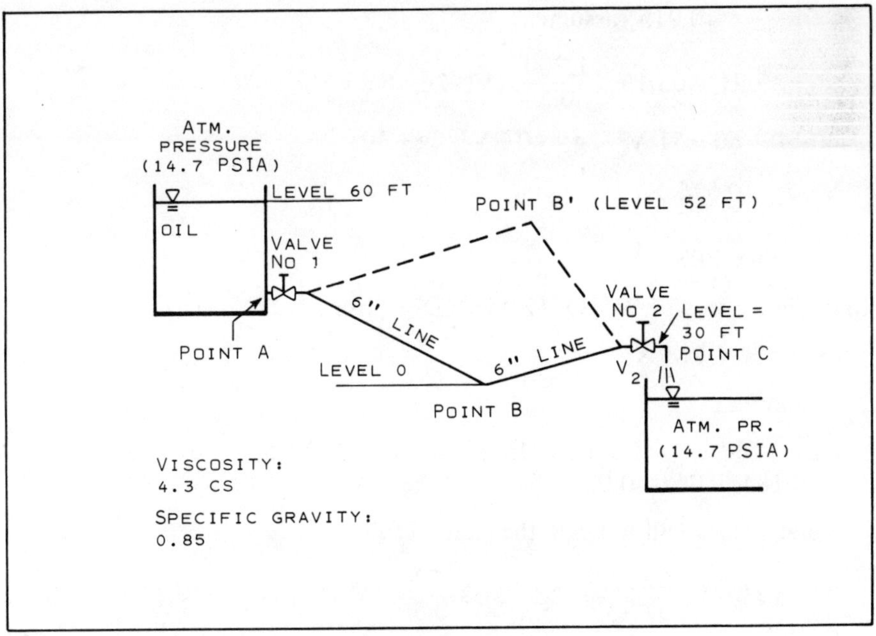

FIG. 3.01

A−B=1,000 FT=A−B′ (alternative route)
B−C= 600 FT=B′−C (alternative route)
One 6″ valve=10 equivalent FT of 6″ steel pipe.
All other losses (entrance, bends, etc)=5% of the total pipe length.
Total equivalent length=1,600+2×10+0.05×1,600=1,700 FT

Solution:

1. $P_1+V_1^2/2g+Z_1-H_f=P_2+V_2^2/2g+Z_2$ (from 3-01)

Where:

$$P_1=14.7\text{ psia}\times\frac{2.31}{0.85}=40\text{ FT}=P_2$$

$V_1=0$

$D=6\text{ in}=0.5\text{ FT}$

$Z_1=60\text{ FT}$

$V_2=$Velocity at pt. C

$Z_2=30\text{ FT}$

$$H_f=f\times\frac{Le}{D}\times V_2^2/2g$$

$f=0.015$ (assumed)

$$H_f=0.015\times\frac{1{,}700}{0.5}\times V_2^2/2g=51\ V_2^2/2g$$

$$40+0+60-51\ V_2^2/2g=40+V_2^2/2g+30$$

$V_2^2/2g=0.575\text{ FT}$

$V_2=6.08\text{ FPS}$

In conjunction with $V=Q/(2.448\times D^2)$ (from 1-22),

Q = 536 GPM Ans.

2. Compute f:

$RE=3162\times Q/(D\times Cs)$ (from 1–24) $=3162\times536/(6\times4.3)=65{,}700$
$f=0.184/RE^{0.2}=0.020$

Note: Tables will indicate the same result.

3. From $H_f=f\times\frac{Le}{D}\times V_2^2/2g=0.02\frac{1700}{0.5}\times V_2^2/2g$, $H_f=68V_2^2/2g$

Using this value in Bernoulli's equation, we obtain:

$V_2^2/2g = 0.43$ FT (as compared to 0.575 FT)

$V_2 = 5.26$ FPS

$Q = 464$ GPM Ans.

This is the flowrate with both valves open and constant liquid levels.

Friction Loss:

$H_f = 68 \times 0.43 = 29.44$ FT (per 1,700 FT) $= 17.20$ FT/1,000 FT

4. Static Head at "B".

Head $= 60 + 14.7 \times 2.31/0.85 = 60 + 40 = 100$ FT ABS $= 36.80$ PSIA $=$ 22.10 PSIG Ans.

5. Energy at "B".

$$P_1 + V_1^2/2g + Z_1 - Hf_{A\text{-}B} = P_B + \frac{V_2^2}{2g} + Z_B$$

where:

$P_1 = 40$ FT (ABS), $V_1 = 0$, $Z_1 = 60$, $Z_B = 0$, and $H_{fA\text{-}B} =$ $0.02 \times \frac{Le_{A\text{-}B}}{D} \times V_2^2/2g = 0.02 \times 1{,}060 \times 0.43 = 9.1$ FT.

Energy at $B = 40 + 0 + 60 - 9.1 = P_B + 0.43 = 90.9$ FT

$P_B = 90.9 - 0.43 = 90.47$ FT (ABS. PRESS. $= 33.29$ psia)

Pressure at B (gauge) $= 33.29 - 14.70 = 18.59$ psig Ans.

Note that a gauge at B will not register the energy, only the pressure. In this case, 18.59 psig.

6. Energy at B′.

Instead of a pipeline run A-B-C, one can have an alternative run A-B′-C, as shown. In this case, the flowrate and losses from A to C would be identical as above. The energy at B′ will be different:

$$P_1 + V_1^2/2g + Z_1 - H_{fA\text{-}B}{}' = P_B{}' + \frac{V_2^2}{2g} + Z_B{}'$$

$40 + 0 + 60 - 9.1 = P_B{}' + 0.43 + 52$

$38.9 =$ ENERGY AT $B' = P_B{}' + 0.43$ (FT ABS)

$P_B{}' = 38.9 - 0.43 = 38.47$ FT ABS $= (38.47/2.31) \times 0.85$ psia $= 14.16$ psia $= -0.54$ psig (partial vacuum) Ans.

Percent Vacuum $= \left(1 - \frac{14.16}{14.70}\right) \times 100 = 3.67 = 1.10$ in. of mercury (partial vacuum).

4.

Compound Pipes in Series

Flow through a pipeline, consisting of sections of different diameters can be computed as follows:

4.1. In case of two different diameters in the line, the friction loss H_f in Bernoulli's Equation 3–01

$$P_1+V_1^2/2g+Z_1-H_f=P_2+V_2^2/2g+Z_2$$

would be equivalent to:

$$H_f=f_A\times\frac{L_A}{D_A}\times\frac{V_A^2}{2g}+f_B\frac{L_B}{D_B}\times\frac{V_B^2}{2g};$$

where:

f_A, f_B=friction factors

L_A, L_B=equivalent pipelengths, FT

V_A, V_B=flow velocities, FPS

D_A, D_B=internal diameters, FT

Index $_A$ refers to first pipe dia. and Index $_B$ refers to second pipe dia.

This relationship can be transformed into:

$$H_f=f_A\frac{L_A}{D_A^5}\times\frac{Q^2}{32.18}+f_B\frac{L_B}{D_B^5}\times\frac{Q^2}{32.18}\text{(from 2–05)}$$

where L is in equivalent feet, D is in inches, and Q is in GPM.

Assume $f_A=f_B=f$. Then:

$$H_f=f\frac{Q^2}{32.18}\left[\frac{L_A}{D_A^5}+\frac{L_B}{D_B^5}\right] \quad \text{(4–01)}$$

And, in general:

$$H_f = f\frac{Q^2}{32.18}\left[\frac{L_1}{D_1^5} + \frac{L_2}{D_2^5} + \ldots + \frac{L_N}{D_N^5}\right] \ldots\ldots\ldots\ldots\ldots\ldots\ldots (4\text{–}02)$$

■ EXAMPLE 4:1

Compute the flowrate through the pipe-system per fig. 4.01, with both valves open and constant liquid levels in both vessels.

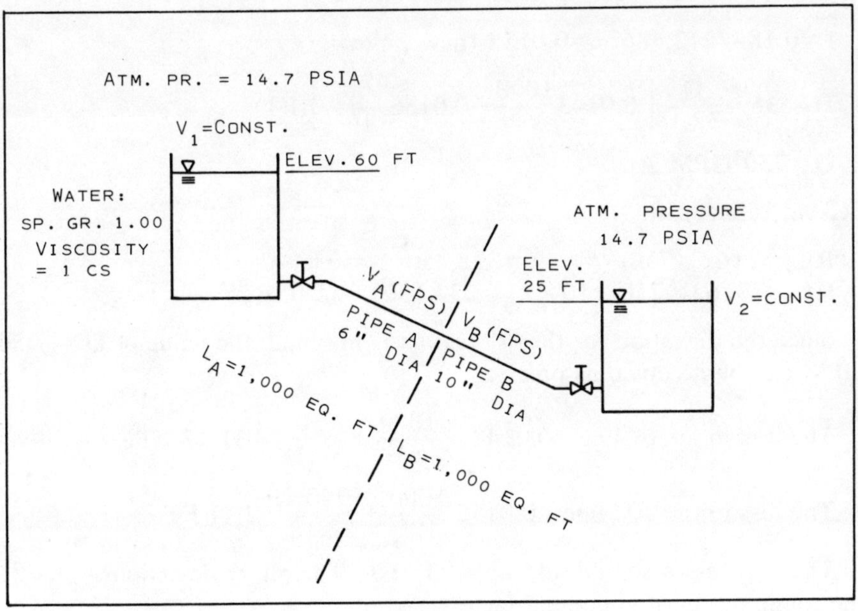

FIG. 4.01

$P_1 + V_1^2/2g + Z_1 - H_f = P_2 + V_2^2/2g + Z_2$ (from 3–01)

$33.96 + 0 + 60 - H_f = 33.96 + 0 + 25$

$$H_f = 35\ \text{FT} = f\frac{Q^2}{32.18}\left[\frac{L_A}{D_A^5} + \frac{L_B}{D_B^5}\right] \text{ (from 4–01)}$$

Assume $f = 0.018$; for $H_f = 35$ FT

$$35 = 0.018\frac{Q^2}{32.18}\left[\frac{1000}{6^5} + \frac{1000}{10^5}\right]$$

$Q = 672$ GPM Ans.

Note: For low-viscosity liquids flowing through new, smooth steel pipes, the "f" factor will range between 0.010–0.020.

For computations not aiming at great precision, the foregoing result can be considered satisfactory. When precision is required, one has to compute RE and f for the 6-in. and the 10-in. lines:

$$RE = 3162 \times GPM/(D'' \times Cs) \text{ (from 1–24)}$$

$$RE\ 6'' = 3162 \times 672/(6 \times 1) = 354{,}144 < 1{,}000{,}000$$

$$f = 0.184/RE^{0.2} \text{ (from 3–02)} = 0.184/354{,}144^{0.2} = 0.0143 \text{ (new pipes)}$$

$$RE\ 10'' = 3162 \times 672/10 \times 1 = 212{,}486$$

$$f = 0.184/212{,}486^{0.2} = 0.0158 \text{ (new pipes)}$$

$$H_f = 35 = \frac{Q^2}{32.18}\left[0.0143\frac{1000}{6^5} + 0.0158\frac{1000}{10^5}\right] \text{(FT)}$$

Q=750 GPM Ans.

Second refinement:

$$RE_6 = 3162 \times 750/(6 \times 1) = 395{,}250;\ f_6 = 0.0140$$
$$RE_{10} = 3162 \times 750/(10 \times 1) = 237{,}150;\ f_{10} = 0.0155$$

Since the deviation for the "f" factors is minimal, the value of Q = 750 GPM can be accepted as precise.

$$\text{The loss in the 6'' line} = 0.0140 \times \frac{1000}{6^5} + \frac{750^2}{32.18} = 31.47 \text{ FT}$$

$$\text{The loss in the 10'' line} = 0.0155\frac{1000}{10^5} + \frac{750^2}{32.18} = 2.71 \text{ FT}$$

The two losses should add up to 35 FT. The slight deviation is due to the rounding-up of computed factors.

The relationship between losses as a function of the different diameters can be established as follows:

$$H_{f1} = f_1\frac{L_1}{D_1{}^5} \times \frac{Q^2}{32.18}$$

$$H_{f2} = f_2\frac{L_2}{D_2{}^5} \times \frac{Q^2}{32.18}$$

$$\frac{H_{f1}}{H_{f2}} = \frac{f_1}{f_2} \times \frac{L_1}{L_2} \times \left(\frac{D_2}{D_1}\right)^5 \text{(for equal Q's)} \quad \ldots\ldots 4\text{–}03$$

for $f_1 \approx f_2$:

$$\frac{H_{f1}}{H_{f2}} \approx \frac{L_1}{L_2}\left(\frac{D_2}{D_1}\right)^5 \text{ (approximate relationship)} \quad \ldots\ldots 4\text{–}04$$

For the above example, in conjunction with 4–03:

$$\frac{H_{f1}}{H_{f2}}=\frac{0.0140}{0.0155}\times\frac{1000}{1000}\times\left(\frac{6}{10}\right)^5=0.07$$

$$H_{f1}+H_{f2}=35=H_f \text{ (from above)}$$

$$H_{f1}=0.07\times H_{f2}$$

$$1.070\ H_{f2}=35$$

$$H_{f2}=32.71\text{ FT}$$

$$H_{f1}=35-32.71=2.29\text{ FT}$$

The results compare favorably with the ones computed above. The slight deviation should not concern the designer since the deviations resulting from the computation of the equivalent lengths and other factors exceed, percentage-wise, the ones above.

4.2. Equivalent length of a given diameter.

Instead of working with two or more diameters, one can introduce an equivalent diameter D_e, its respective equivalent length L_e, and friction factor f_e, for which the friction loss would be identical.

$$\text{If } H_f=f\frac{L}{D^5}\times\frac{Q^2}{32.18}$$

and

$$H_{fe}=f_e\frac{L_e}{D_e^5}\times\frac{Q^2}{32.18}$$

then, for $H_f=H_{fe}$,

$$\frac{L_e}{L}=\frac{f}{f_e}\times\left(\frac{D_e}{D}\right)^5 \text{ (for equal Q's)} \qquad \text{4–05}$$

■ EXAMPLE 4:2

Compute the flow per fig. 4.01, using equivalent length. Express the 10 in. line in terms of 6 in. ($f_{10}''=0.0155$; $f_6''=0.0140$):

$$L_e=L\times\frac{f}{f_e}\times\left(\frac{D_e}{D}\right)^5=1000\times\frac{0.0155}{0.0140}\times\left(\frac{6}{10}\right)^5=86.$$

Thus:

$$L_6''=1{,}000+86=1{,}086\text{ FT}$$

$$H_f=35\text{ FT}=0.0140\times\frac{1086}{6^5}\times\frac{Q^2}{32.18}$$

$$Q=758\text{ GPM Ans.}$$

4.3. Replacing one pipeline with n equivalent pipelines, or vice versa.

Design Criterion: The pressure loss through the lines will be identical:

$$H_f = f_1 \frac{L}{D_1^5} \times \frac{Q^2}{32.18} = f_2 \frac{L}{D_2^2} \times \frac{(Q/n)^2}{32.18}$$

For equal lengths and "f" factors:

$\frac{Q^2}{D_1^5} = \frac{(Q/n)^2}{D_2^5}$; and $n^2 D_2^5 = D_1^5$ (for equal Q's)

With n known by requirement:

$$D_2 = D_1/n^{0.4} \qquad \text{4–06}$$

Note: For a standard size selection, this equation could be considered satisfactory. A refinement could be achieved by means of computing the "f" factors and adjusting accordingly. (For different "f" factors 4–06 transforms into $f_1/D_1^5 = f_2/D_2^5 n^2$).

■ EXAMPLE 4:3

A 60″ pipe (60″ OD; 59″ ID) has to be replaced by three pipes of equal diameter (0.5″ wall thickness).

$D_2 = D_1/n^{0.4}$ (from 4–06)

$D_2 = 59/3^{0.4} = 38''$ ID.

The equivalent size of each of the three pipes would be 39″ OD.

In order to refine the result (if required), one has to compute the "f" factors. Assume a flowrate corresponding to 8 FPS:

Q = 68,000 GPM (60″ pipe);

$RE_{60''} = 3162 \times 68{,}000/(59 \times 1) = 3{,}644{,}000$

From tables: f = 0.011

$$RE_{39''} = \frac{3162 \times 68000/3}{38 \times 1.0} = 1{,}886{,}000$$

$f_{39''} = 0.012$

Then: $\frac{0.011}{59^5} = \frac{0.012}{D_2^5 \times 3^2}$

$D_2 = 38.68''$

In this case, the equivalent size would be 39″. Depending on an economic analysis, a 39″ OD, or 40″ OD would be selected. Ans.

■ EXAMPLE 4:4

Four 8″ (7.981″ ID) pipes have to be replaced by one equivalent line. Consider equal "f" factors:

$D_2 = D_1/n^{0.4}$ (from 4–06)

$D_1 = D_2 \times n^{0.4} = 7.981 \times 4^{0.4} = 13.89''$ ID

Select a 14″ OD or 16″ OD line upon economic evaluation. Ans.

5.

Graphical Representation of Friction Losses in a Piping System

As pointed out:

$$H_f = f\frac{L_e}{D^5} \times \frac{Q^2}{32.18} \text{ (from 2–05)}$$

When computing the pure friction loss in pipes, L is meant to represent the physical length of the pipe. When including exit, entrance losses, and losses through equipment, bends, and other fittings, L_e, as indicated before, represents the equivalent length of the system.

In case the pipeline is long, L will approach the value of L_e. For preliminary computations one can assume that Le = 1.10 L. On the other hand, when the pipes are short, like around a pumping station or in a refinery, the friction loss through the equipment and fittings will, by far, exceed the pipe friction loss. In this case a detailed listing of the separate equipment items will be required, and their "K" factors, or equivalent lengths determined.

For a given Le, f, and D, the above formula can be expressed simply:

$$H_f = C \times Q^2 \text{ (FT)} \quad \ldots\ldots 5\text{–}01$$

where

$$C = f\frac{Le}{D^5} \times \frac{1}{32.18} \quad \ldots\ldots 5\text{–}02$$

This is a parabola, called Friction Loss System Curve. It can be represented graphically, as shown on fig. 5.01. The curve is also called "System friction curve," "Friction loss curve," or "Friction head curve."

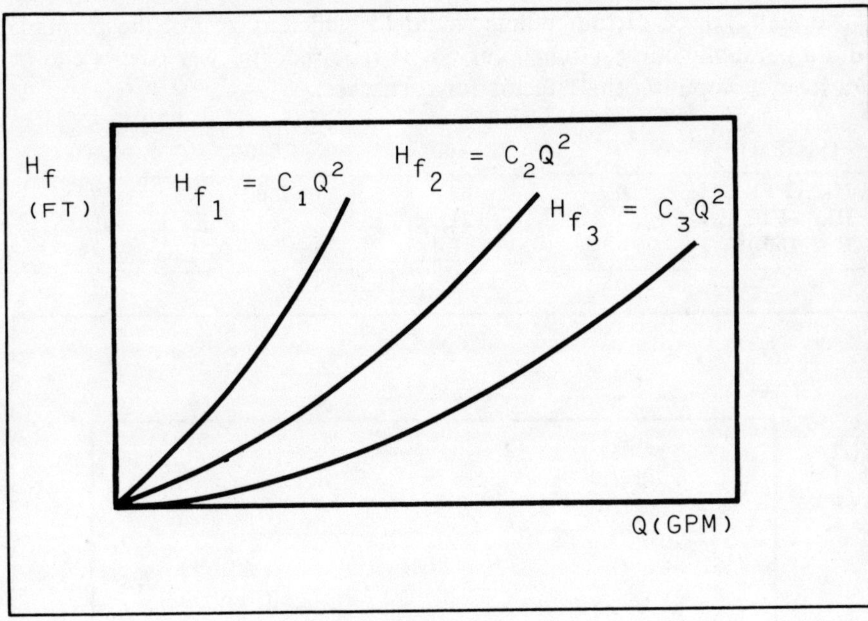

FIG. 5.01
Friction loss system curves.

■ EXAMPLE 5:1

Plot the friction loss system curves for the following conditions:

f_1, f_2, f_3 = 0.018; Le = 15,000 FT; D_1 = 8 in, D_2 = 10 in, and D_3 = 12 in.

Solution:

Compute C per 5–02.

$C = f\,Le/(D^5 \times 32.18)$

$C_{8''} = 0.018 \times 15{,}000/(8^5 \times 32.18) = 1/3{,}905$

$C_{10''} = 0.018 \times 15{,}000/(10^5 \times 32.18) = 1/11{,}918$

$C_{12''} = 0.018 \times 15{,}000/(12^5 \times 32.18) = 1/29{,}657$

Thus:

$H_{f8''} = Q^2/3{,}905$ (FT)

$H_{f10''} = Q^2/11{,}918$ (FT)

$H_{f12''} = Q^2/29{,}657$ (FT)

Normally, three or four points would be sufficient to plot the parabola with a parabolic curve (french curve). If required, preciser curves can be drawn by computing the f factor for each case.

Q (GPM)	0	500	1,000	1,500
H_{f8}" (FT)	0	64	256	576
H_{f10}" (FT)	0	21	84	188
H_{f12}" (FT)	0	8.4	34	76

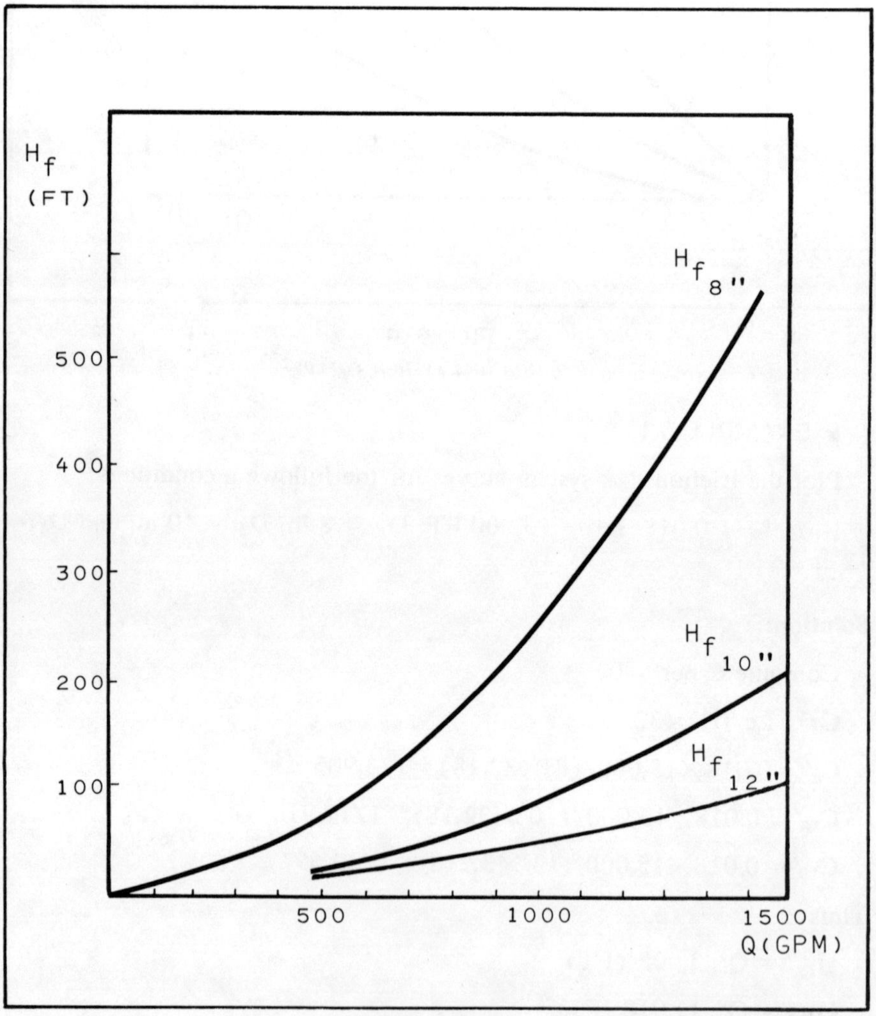

FIG. 5.02
Friction loss system curves.

When changing flowrate in an existing system:

$H_{f1}=C\times Q_1^2$, and $H_{f2}=C\times Q_2^2$

$H_{f1}/H_{f2}=(Q_1/Q_2)^2$ 5–03

5–03 assumes an identical "f" factor for both cases. This relationship is satisfactory for preliminary computations.

6. Compound Pipes in Parallel

Flowrates in two or more parallel pipes, branching off at a common point, are computed based on the existing head (hydraulic gradient) at it. This head is common to all branches.

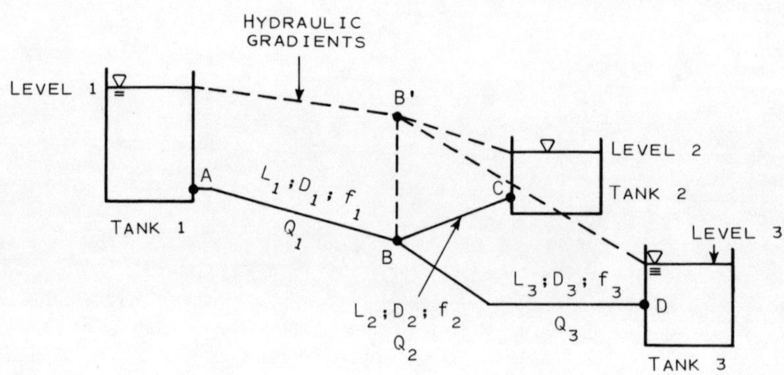

FIG. 6.01

In the case per fig. 6.01, B′ represents the hydraulic gradient, or head at point B. The flowrate from B to tank 2 will be based on the differential B′ − Level 2, and the one to tank 3 on the differential B′ − Level 3.

The main flowrate from tank 1 to B will be based on the differential Level 1 − B′.

From 2–05:

6.1. Level $1 - B' = H_{f1} = f_1 \times L_1 \times Q_1^2/32.18\ D_1^5$

6.2. $B' -$ Level $2 = Hf_2 = f_2 \times L_2 \times Q_2^2/32.18\ D_2^5$

6.3. $B' -$ Level $3 = Hf_3 = f_3 \times L_3 \times Q_3^2/32.18\ D_3^5$

6.4. $Q_1 = Q_2 + Q_3$

where:

f = friction factor, L = equivalent length in FT, Q = flowrate (GPM), D = internal dia. (inches), and H_f = friction loss in FT.

In the above equations, all parameters, with the exception of B′ and the three flowrates, are known. One assumes a hydraulic grade B′ and resolves for all Q's. If the assumption proves correct, equation 6.4 will be satisfied. If not, a second, and may be a third trial, will be required.

■ EXAMPLE 6:1

Refer to fig. 6.01. Compute flowrates Q_1, Q_2, Q_3 for the following conditions:

Level 1, 2, 3 = 80 FT, 50 FT, 0 FT

L_1, L_2, L_3 = 2,000 FT, 3,000 FT, 4,000 FT

D_1, D_2, D_3 = 24 in, 12 in, 16 in

f_1, f_2, f_3 = 0.025, 0.0185, 0.017

$$6.1.\ 80 - B' = 0.0250 \frac{2000}{24^5} \times \frac{Q_1^2}{32.18} = Hf_1 = Q_1^2/5{,}125{,}000$$

$$6.2.\ B' - 50 = 0.0185 \frac{3000 \times Q_2^2}{12^5 \times 32.18} = Hf_2 = Q_2^2/144{,}280$$

$$6.3.\ B' - 0 = 0.0170 \frac{4000 \times Q_3^2}{16^5 \times 32.18} = Hf_3 = Q_3^2/496{,}220$$

6.4. $Q_1 = Q_2 + Q_3$

Assume $B' = 70$; Then:

6.1. $Hf_1 = 80 - 70 = 10 = Q_1^2/5{,}125{,}000$; $Q_1 = 7{,}158$ GPM

6.2. $Hf_2 = 70 - 50 = 20 = Q_2^2/144{,}280$; $Q_2 = 1{,}700$ GPM

6.3. $Hf_3 = 70 - 0 = 70 = Q_3^2/496{,}220$; $Q_3 = 5{,}893$ GPM

$1{,}700 + 5{,}893 = 7{,}593 > 7{,}158$

Assume $B' = 69$ FT, then:

6.1. $Hf_1 = 80 - 69 = 11 = Q_1^2/5{,}125{,}000$; $Q_1 = 7{,}508$ GPM

6.2. $Hf_2 = 69 - 50 = 19 = Q_2^2/144{,}280$; $Q_2 = 1{,}655$ GPM

6.3. $Hf_3 = 69 - 0 = 69 = Q_3^2/496{,}220$; $Q_3 = 5{,}851$ GPM

$1{,}655 + 5{,}851 = 7{,}506 \approx 7{,}508$ GPM

$Q_1 = 7{,}508$ GPM; $Q_2 = 1{,}655$ GPM; $Q_3 = 5{,}851$ GPM

$B' = 69$ FT Ans.

In case tank 2 and tank 3 in fig. 6.01 have the same liquid levels, as indicated in fig. 6.02, then B' − level 2 = B' − level 3, or $Hf_2 = Hf_3$.

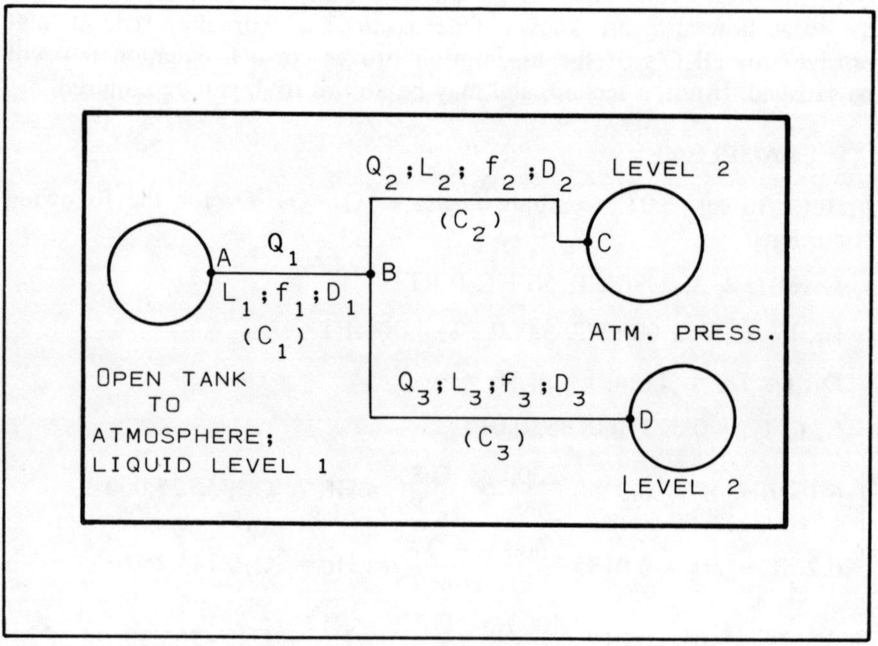

FIG. 6.02

6.5. Head loss in BC = head loss in BD

6.6. $Q_1 = Q_2 + Q_3$

6.7. $Hf_1 = f_1 \dfrac{L_1}{D_1^5} \times \dfrac{Q_1^2}{32.18}$

6.8. $Hf_2 = Hf_3 = f_2 \frac{L_2}{D_2^5} \times \frac{Q_2^2}{32.18} = f_3 \frac{L_3}{D_3^5} \times \frac{Q_3^2}{32.18}$

6.9. $Hf_1 + Hf_2$ (or Hf_3) = Level 2 − Level 1 (in feet)

Normally, only flowrates and the "f" factors are unknown. It is accepted to assume for starters that all "f" factors are equal. This will facilitate the computation of a first value for the capacities. Thereafter, the computation is refined by computing the "f" factors for each branch, and recomputing the capacities.

The analytical method can sometimes become tedious, especially when precision is required. The graphical solution can efficiently provide faster solutions. Refer to fig. 6–03.

The above equations can be simplified by expressing them as follows:

6.6. $Q_1 = Q_2 + Q_3$

6.7. $Hf_1 = C_1 \times Q_1^2$

6.8. $Hf_2 = C_2 \times Q_2^2$; $Hf_3 = C_3 Q_3^2$

6.9. $Hf_1 + Hf_2 = K$ (given constant) = level 1 − level 2.

Plot the system friction curves (Fig. 6.03) for:

$Hf_1 = C_1 Q^2$; $Hf_2 = C_2 Q^2$; $Hf_3 = C_3 Q^2$

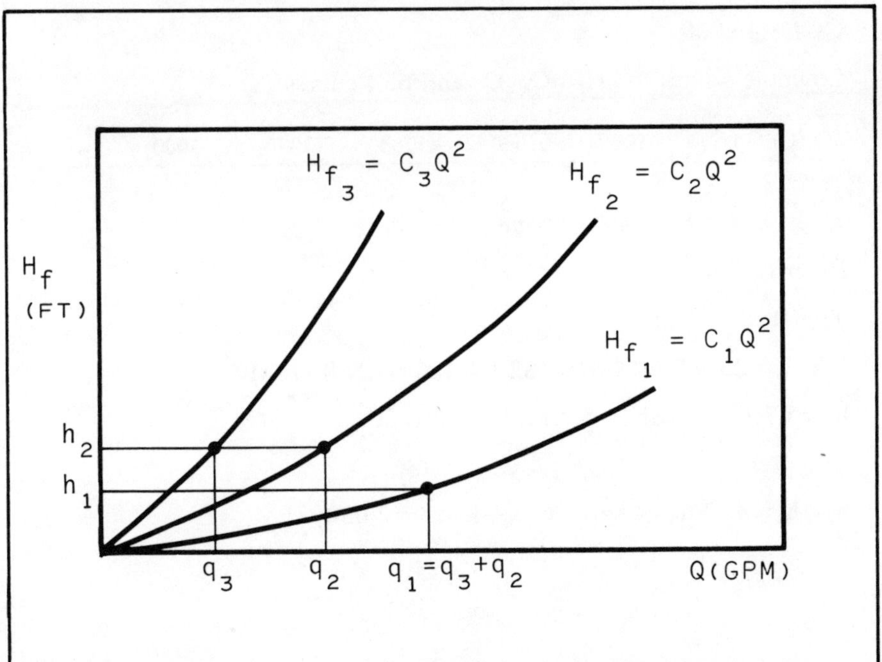

FIG. 6.03

H_f values on curves H_{f2} and H_{f3} will be identical only for the capacity resulting from the specific conditions outlined. Draw a tentative line at h_2, to cross Hf_2 and Hf_3. Read off the values q_3 and q_2 and the corresponding loss h_1 for a capacity equal to $q_3+q_2=q_1$ from curve Hf_1. Then check whether equation 6.9 above is satisfied. In case it is, the flow rates have been determined. Normally, two or three trials will help obtain the answer.

■ EXAMPLE 6:2

Determine flow conditions for the following data (fig. 6.02):

Level 1=60 feet

Level 2=10 feet

Identical "f" factors=0.018 (steel pipes)

$L_1=1{,}000$ FT, $ID_1=12$ in, $L_2=1{,}500$ FT, $ID_2=6$ in., $L_3=1{,}200$ FT, and $ID_3=8$ in.

$Hf_1+Hf_2=60-10=50$ ft

$C=f\frac{L_e}{D^5}\times\frac{1}{32.18}$ (per 5–02)

$C_1=1/445{,}000$

$C_2=1/9{,}300$

$C_3=1/49{,}000$

Compute values for $H_f=C\times Q^2$ and plot curves.

Q (GPM)	500	1000	1500	2000	3000	4000
H_{f1} (FT)			5	9	20	36
H_{f2} (FT)	26	108	242			
H_{f3} (FT)	5	20	46	82		

Refer to Fig. 6.04.

Trial 1:

$h_2=h_3=25$ FT; $q_2+q_3=482+1{,}106=1{,}588$ GPM

$h_1=5.7$ FT; $h_1+h_2=5.7+25=32.7\neq50$.

Trial 2:

$h_2=h_3=41$ FT; $q_2+q_3=615+1{,}415=2{,}030$ GPM

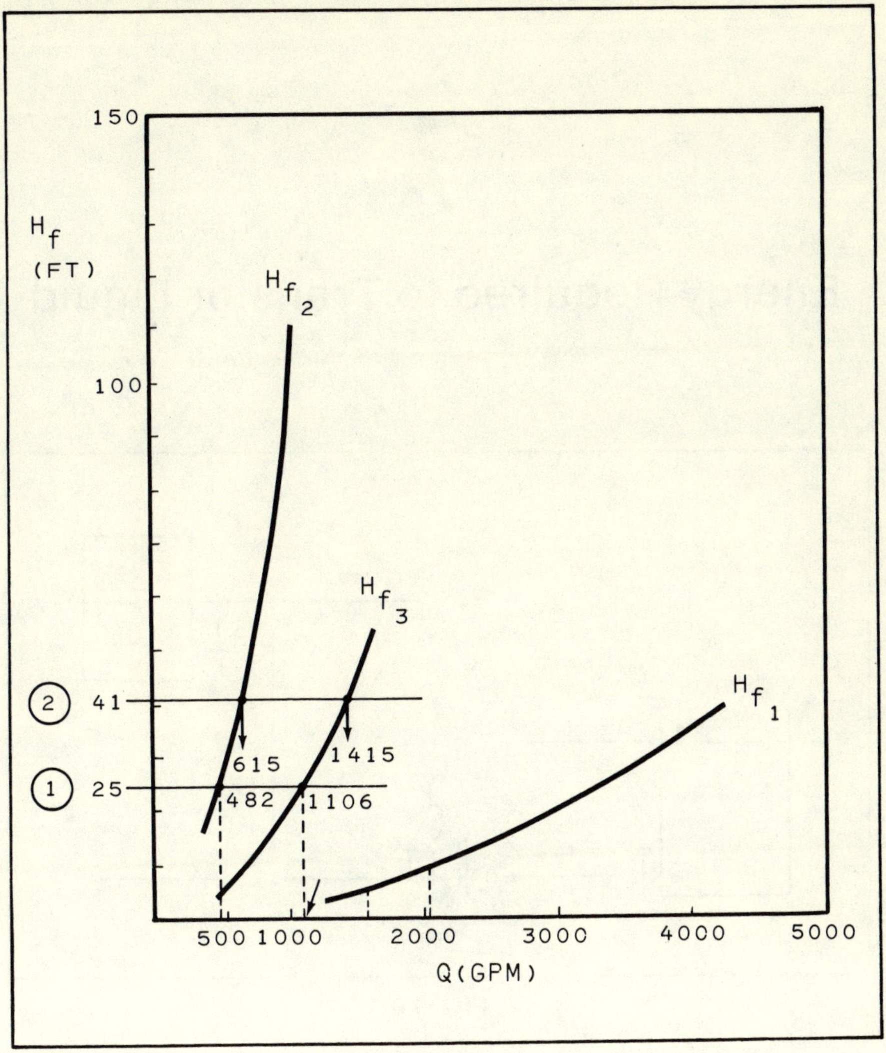

FIG. 6.04

$h_1 = 9$ FT; $h_1 + h_2 = 9 + 41 = 50$ FT−OK.

$Q_1 = 2{,}030$ GPM

$Q_2 = 615$ GPM

$Q_3 = 1{,}415$ GPM Ans.

Refinements of the above answers have to be done analytically by computing "RE" and "f".

7.

Energy Required to Transfer Liquid

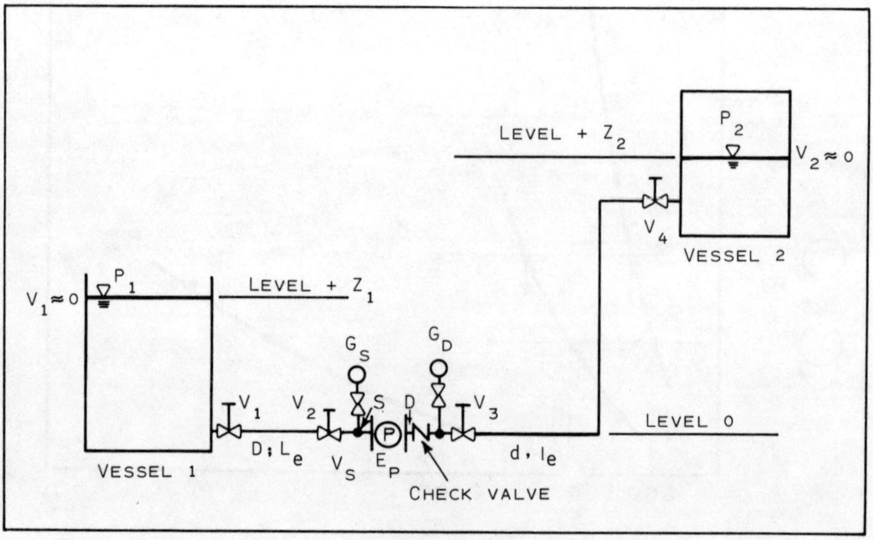

FIG. 7.01

Assume the requirement is to transfer Q GPM of a liquid from vessel 1 to vessel 2 through the system shown on fig. 7–01. The energy for the tranfer (E_p) will be provided by a pump.

From Bernoulli's equation (3–01):

$$P_1 + Z_1 + V_1^2/2g + Ep - H_f = P_2 + Z_2 + V_2^2/2g$$

where:

$$H_f = f_1 \frac{L_e}{D^5} \times Q^2/32.18 + f_2 \frac{le}{d^5} \times Q^2/32.18 = \frac{Q^2}{32.18}\left[f_1 \frac{L_e}{D^5} + f_2 \frac{le}{d^5} \right]$$

= sum of all losses, from vessel 1 to pump, and from pump to vessel 2.

$$P_1+Z_1+O+E_p-\frac{Q^2}{32.18}\left[f_1\frac{L_e}{D^5}+f_2\frac{le}{d^5}\right]=P_2+Z_2+0$$

$$E_p=(P_2-P_1)+(Z_2-Z_1)+\frac{Q^2}{32.18}\left[f_1\frac{L_e}{D^5}+f_2\frac{le}{d^5}\right](FT) \quad \ldots\ldots\ldots\ldots 7\text{–}01$$

Comment: Elevation Zero (0) is assigned to the pump centerline. All elevations have to be added algebraically, i.e., with their signs. It should be noted that E_p is equal to the energy differential between points D and S. Refer to example 7:1.

■ EXAMPLE 7:1

Refer to Fig. 7–01. Transfer 2,000 GPM of water at 60° F (sp. gr. = 1.00).

P_1 = 14.7 psia = 34 FT

Z_1 = 40 FT

L_e = 150 FT (suction side)

D = 12″ (internal diameter)

d = 10″ (internal diameter)

le = 1,500 FT (discharge side)

P_2 = 100 psig = 114.7 psia = 265 FT. ABS.

Z_2 = 80 FT

$f_1 = f_2 = 0.02$ (not a new pipe)

7.1 Energy required:

$$E_p=(P_2-P_1)+(Z_2-Z_1)+\frac{Q^2}{32.18}\left[f_1\frac{L_e}{D^5}+f_2\frac{le}{d^5}\right](\text{from } 7\text{–}01)$$

$$=(265-34)+(80-40)+\frac{2000^2}{32.18}\left[0.02\frac{150}{12^5}+0.02\frac{1500}{10^5}\right]$$

$$=231+40+39;\ (H_{f\ suction}=1.5;\ H_{f\ disch.}=37.5)$$

E_p = 310 FT of water = 310/2.31 = 134.2 psi Ans.

This energy is referred to as TOTAL HEAD, or TOTAL DIFFERENTIAL HEAD. It is the energy imparted by the pump for the required transfer.

7.2. Compute energy levels at points S and D during the transfer of 2,000 GPM of water.

Per Bernoulli, at S:

$$P_1+Z_1+V_1^2/2g-H_{fs}=P_s+Z_s+V_s^2/2g$$

Where:

$H_{fs}=1.5$ FT (per 7.1)
$V_s=2{,}000/(2.448\times12^2)=5.67$ FPS (1–22)
$V_s^2/2g=0.5$ FT
$34+40+0-1.5=P_s+O+0.5$
$P_s+0.5=72.5$ FT ABS $=31.38$ PSIA $=$ energy level at S

Note: A gauge at S would not read the velocity head. Hence, the gauge would indicate 16.46 psig, the pressure at S.

($P_s=72.5-0.5=72$ FT ABS $=31.16$ PSIA $=16.46$ PSIG)

Conclusion: Add velocity head to gauge reading, to obtain precise energy level (dynamic suction pressure) at S.

Per Bernoulli, at D:

$$P_1+Z_1+V_1^2/2g+E_p-H_{fs}=P_D+Z_D+V_D^2/2g$$

Where:

$H_{fs}=1.5$ FT (per 7.1)
$V_D=2000/(2.448\times10^2)=8.17$ FPS
$V_D^2/2g=1.04$ FT
$E_p=310$ FT (from 7.1)
$34+40+0+310-1.5=P_D+O+1.04$
$P_D+1.04=$ Energy at D $=382.5$ FT ABS $=165.58$ PSIA

Note: A gauge at D would not read the velocity head. Hence the gauge would indicate 150.43 psig, the pressure at D.

($P_D=382.5-1.04=381.46$ FT ABS $=165.13$ PSIA $=150.43$ PSIG)

Conclusion: Add velocity head to gauge reading, in order to obtain precise energy level (dynamic discharge pressure) at D.

Energy differential = total head = pump head = $(P_D+V_D^2/2g)-(P_s+V_s^2/2g)=382.5-72.5=310$ FT as in 7.1

Note: Subtracting gauge pressures across the pump can deviate from the true Total Head by 0.1% – 10%, depending on the pipe diameters and pressures. In this case, $150.43-16.46=133.97$ psig $=309.47$ FT, a deviation of 0.17%.

Pressures at points S and D are measured in PSIA or PSIG, and are called suction and discharge pressures. When measured in feet, they are called suction and discharge heads. (Note: They are in feet of liquid pumped).

The TOTAL HEAD represents the total work done by the pump per one pound of liquid pumped. It includes the friction losses, as well as the energy required to overcome pressure and level differentials. Work = work/lb of liquid pumped = ft-lbs/lb=feet of liquid.

We can represent equation 7–01 graphically. In it, H_f represents all the friction losses from point 1 to point 2. It is a parabola of the form $C \times Q^2$. E_p, also indicated as H, is the ordinate in the coordinate system and Q−the abscissa. The units: $E_p = H$, in feet of liquid pumped, and Q_1 is in GPM.

Fig. 7.02 shows H−Curves, their distance away from "O" at Q=O, reflecting the pressure and/or level differentials. These curves are also called TOTAL SYSTEM-HEAD CURVES, or simply SYSTEM-HEAD CURVES. Fig. 7.03 indicates physical layouts of pumping systems, corresponding to the curves in 7.02.

The careful study and thorough understanding of the total-system-head curves, and the way they relate to the physical layouts are indispensable for further understanding of the chapters to follow.

Note that in example 3a (fig. 7.02), we added the velocity head at the exit of the pipe. This follows from Bernoulli's equation 3–01. In the other cases, the velocity heads at the initial and final points are almost zero, since the velocity of the liquid levels in the tanks is minimal during the pumping operation.

■ EXAMPLE 7:2

Determine pump energy required for case per Fig 7.04.

Pressure differential$=P_2-P_1$ (in feet ABS), static elevation differential $=+Z_2-(-Z_1)$ (FT), and total friction loss$=H_f$, in feet. Then:

$E_p = H_f + (P_2 - P_1) + (Z_2 - Z_1)$

for $P_2 = P_1 =$ atmospheric

$Z_2 = +60$ FT and $Z_1 = -10$ FT

$E_p = H_f + [60 - (-10 \text{ FT})] = H_f + 70$, in FT Ans.

■ EXAMPLE 7:3

Determine Ep for the arrangement per fig 7.05.

$E_p = H_f + [Z_2 - Z_1]$

For $Z_1 = -10$ FT; $Z_2 = -30$ FT; $P_2 = P_1$

$E_p = H_f + [-30 - (-10)] = H_f - 20$ Ans.

The static elevation differential is −20 FT. This indicates that, without pump energy, there will be flow by gravity which can be computed from the relationship $H_f = f\frac{L_e}{D^5} \times \frac{Q_s^2}{32.18} = 20$.

For a flowrate larger than Q_s, a pump will be required.

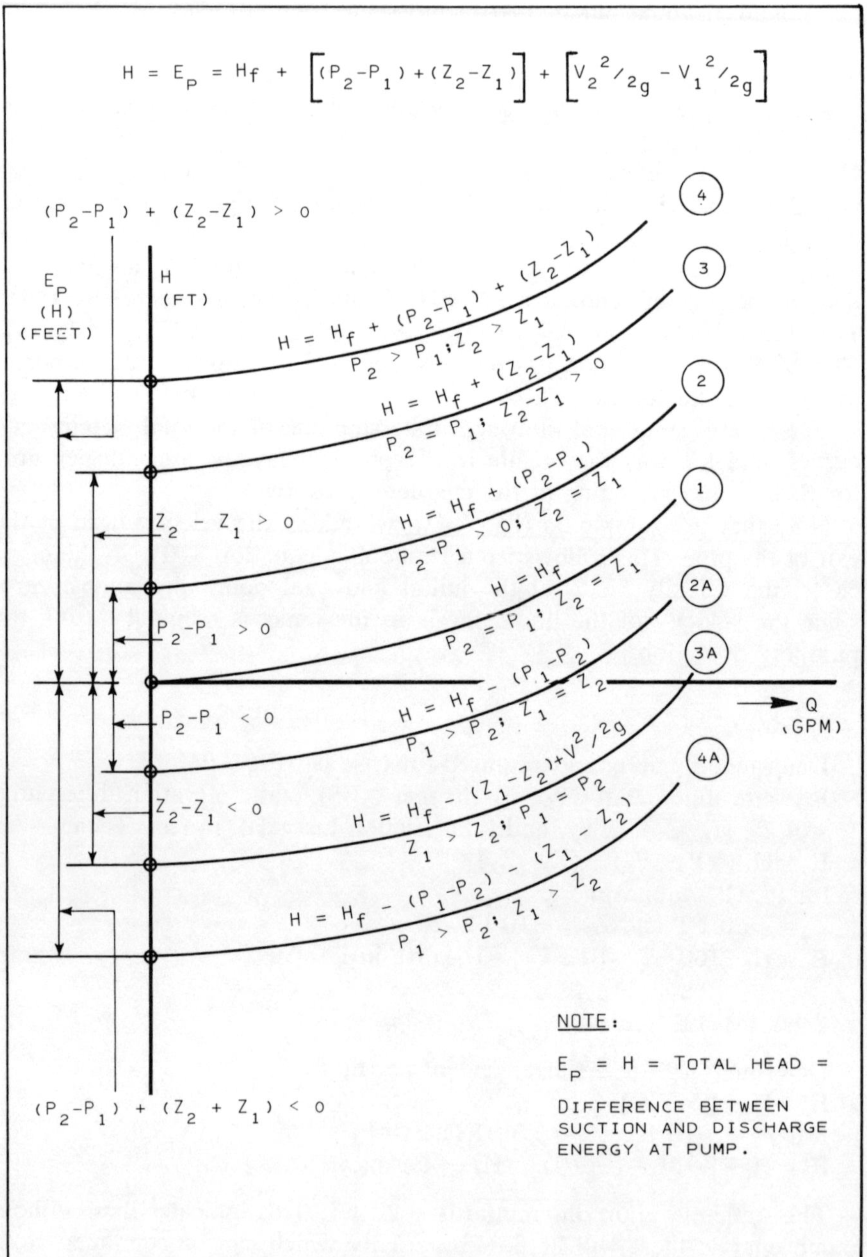

FIG. 7.02

Total system head curves for various pressure and level differentials.

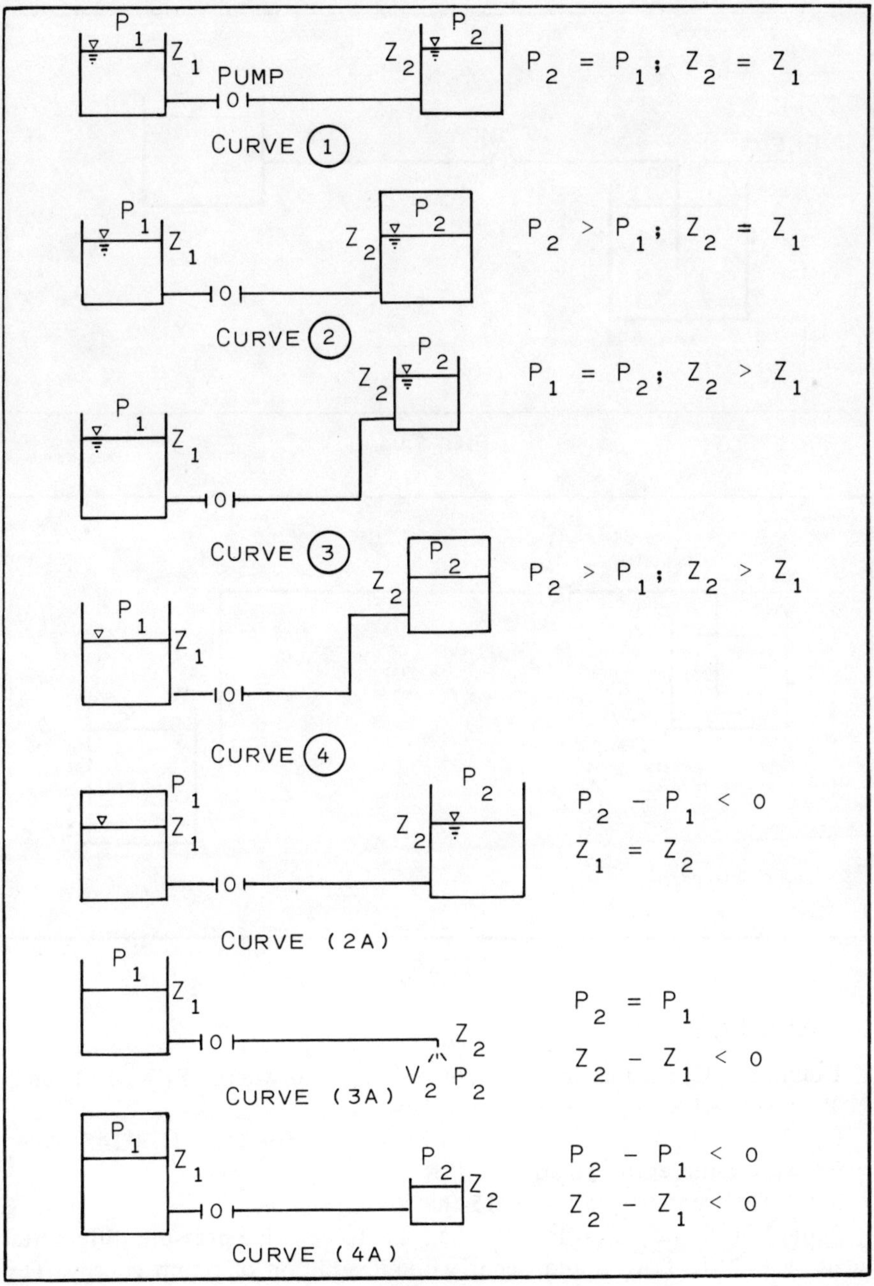

FIG. 7.03

Typical pumping system layouts represented graphically by system curves in fig 7.02

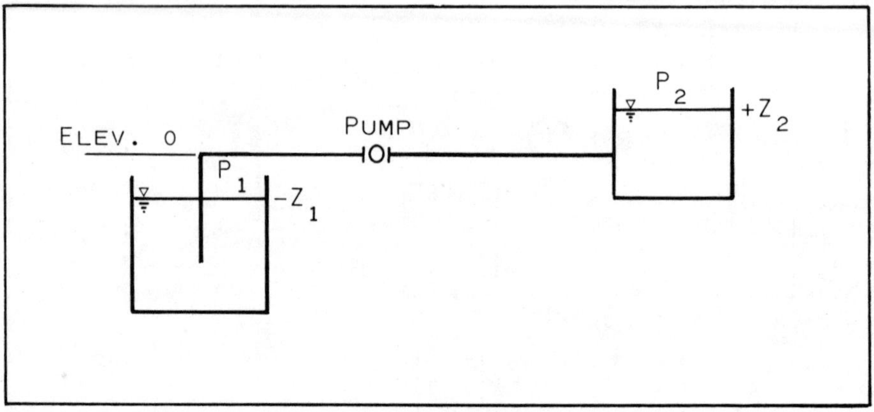

FIG. 7.04

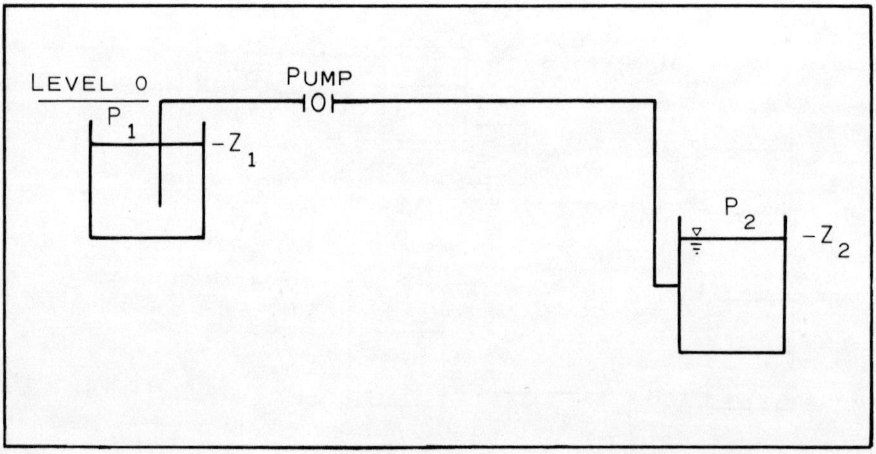

FIG. 7.05

▪ EXAMPLE 7:4

Determine E_p for the arrangement per fig. 7.06 where: $P_1=100$ ft abs., $P_2=30$ ft ABS., $Z_1=-15$ ft, and $Z_2=+200$ ft.
$E_p=H_f+(30-100)+(200-(-15))=H_f-70+215=H_f+145$ Ans.
Pressure Differential Head $=-70$ ft
Static Differential Head $=+215$ ft

Note: For $Z_2=Z_1$; $E_p=H_f-70$, i.e., due to negative pressure differential head of -70 ft, flow would occur without addition of pump energy. The flow rate can be computed from: $H_f=70=f\frac{L_e}{D^5}\times\frac{Q_s^2}{32.18}$

For flowrates higher than Q_s, pump energy will be required.

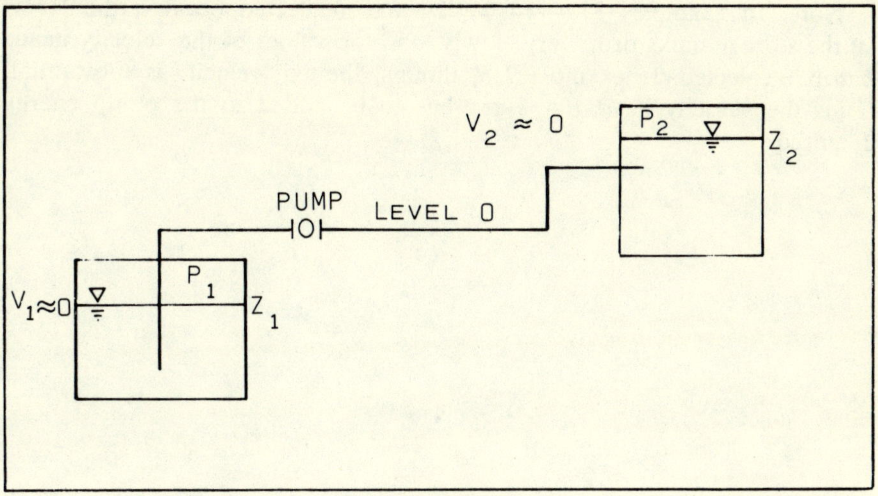

FIG. 7.06

■ EXAMPLE 7:5

Determine E_p per fig. 7.07.

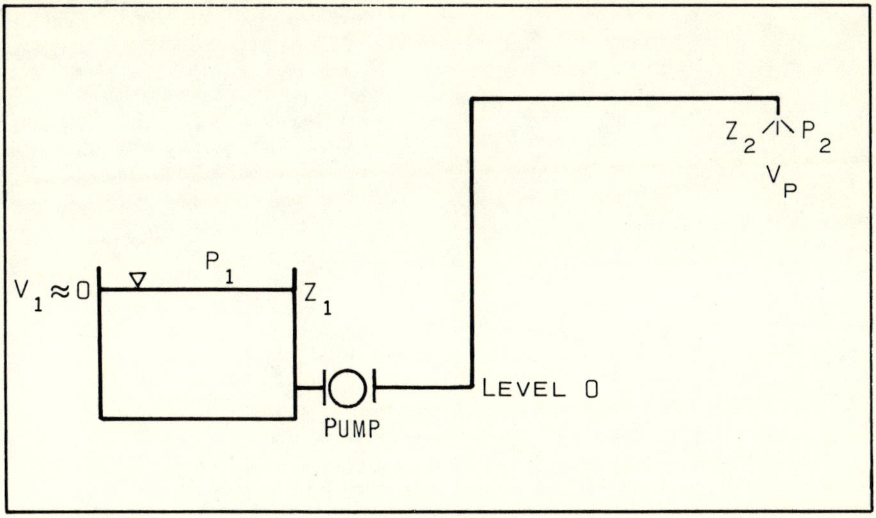

FIG. 7.07

$E_p = (P_2 - P_1) + (Z_2 - Z_1) + H_f + V_p^2/2g = (Z_2 - Z_1) + H_f + V_p^2/2g$
For $Z_2 = 50$ FT; $Z_1 = 20$ FT; $V_p = 8$ FPS, $V_p^2/2g = 1$ FT,
$E_p = H_f + 30 + 1 = H_f + 31$ Ans.

Note: In examples 7:2–7:4 $V^2/2g$ was neglected because the levels in the storage tanks drop very slowly, and the values of the velocity heads can be neglected. In example 7:5, though, the exit velocity is substantial. Thus the velocity head (= 1 ft) has to be added to the pump energy required.

8.

The Centrifugal Pump

8.1. The centrifugal pump is a machine which provides the energy required to transfer liquids from one point to another. Chapter 7 shows how to compute this energy. While a pump is in operation and providing the energy required, one can measure the pressures generated by means of gauges at any point of the system, or compute them as shown in example 7:1.

Due to the incompressibility of liquids (true for practical purposes), all energy imparted to a liquid by the pumping action tends to put it in motion. In case a closed valve will not allow motion, pressure, called "shut-off," will build up. Since there is no motion, the entrapped liquid will heat up, and eventually partial evaporation may take place, a condition which must be avoided.

Upon pump shut-down (refer to figure 7–01), with tank valve V_4 and pump discharge valve V_3 open, the liquid will stop moving (no energy imparted), and the pressure will reverse from dynamic to static. The check valve in the discharge line prevents backflow. Gauge G_D will indicate $Z_2/2.31$ psig for water. With valve V_3 closed, and V_1 and V_2 open, gauge G_s will indicate $Z_1/2.31$ psig for water.

In case the pump is left running at shut-off with V_2 and V_3 open, and V_4 closed, gauge G_s, with valve V_1 open, will indicate $Z_1/2.31$ psig, and gauge G_D will indicate pump shut-off pressure. Upon pump shut-down, the shut-off pressure will continue to prevail in the discharge line between the check valve and valve V_4. With time, though, due to imperfect seals, the pressure will start dropping. In case this pressure has to be maintained (as in fire water lines), the pump can be started at shut-off pressure for a very short time period, and it will be restored. On the other hand, upon opening valve V_4, while the pump is idle, the shut-off pressure in the line will drop to $Z_2/2.31$. This will be the pressure indicated by gauge G_D.

The centrifugal pump, horizontal or vertical, has a casing or bowl in which one or more impellers are mounted on a shaft. Accordingly, the pump is classified as single or multistage. The shaft rotates at a specified speed (rotations per minute=RPM). The rotating impellers impart the energy to the liquid to be transferred by means of centrifugal force.

For construction and classification details of the many pump types the reader is advised to refer to textbooks and manufacturers' catalogs.

The pump is driven by a driver (electric motor, steam or gas turbine, diesel engine, gas or gasoline engine, steam engine). Most centrifugal pumps are electric-motor driven, in most of the cases directly connected to the pump, at speeds of 1,200 1,750, or 3,600 RPM.

The energy imparted by the centrifugal pump is a function of the impeller diameter and its rotating speed. The rate of transfer (flowrate, capacity) for a given pump varies between limits established by the size and construction of the pump.

When computing the energy expense for the transfer of a liquid at a given rate from one point to another, one has to consider the losses in the pump and in the driver. Depending on the conditions, a pump can operate at efficiencies ranging from 20 to 90 percent. The efficiencies are indicated by the manufacturer.

A pump which is 75% efficient in transfering a liquid at the rate of 2,000 GPM per example 7:1 requires 310/0.75=413.33 FT of water, 310 FT for the transfer itself, and 103.33 FT to overcome the internal losses of the pump.

The Horsepower required (HP) is represented by the total work done per unit time in transferring the 2,000 GPM (FT-LBS/unit time) against a head of 310 FT.

2,000 GPM=267.4×62.33 (see 1-14) lbs/min=16,667 lbs/min.
[2,000 GPM=267.4 CFM (Table 1.03)]

Work done=16,667×310 ft=5,166,770 ft-lbs/min=156.57 H.P.
(1 HP=33,000 ft-lbs/min, per 1-16).
156.57/310=0.505 HP/FT.
In the same line, we require for the pump losses:
103.33 FT × 0.505 HP/FT=52.18 HP.
Total Pump HP required=156.57+52.18=208.75.

The horsepower requirement at the pump shaft is called BRAKE HORSEPOWER (BHP). In this case, BHP=208.75.

The motor HP will be the next higher standard size, in this case 250 HP. Driver efficiencies vary. (Electric motors: 93-97%).

$$BHP=\frac{LB-FT}{MIN}\times\frac{1}{33,000}$$

In general: (see table 1.03)

$$\frac{Q(GPM)\times 0.1337\times 62.33\times Sp.g.\times Total\ Head\ (FT)}{33{,}000\times Pump\ Efficiency\ (percent,\ fraction)}=BHP \quad \ldots\ldots 8\text{–}01$$

$$BHP=\frac{Q(GPM)\times H\ (FT)\times Sp.g.\ of\ liquid}{3{,}960\times Pump\ Efficiency\ (percent,\ fraction)} \quad \ldots\ldots\ldots\ldots\ldots 8\text{–}02$$

The driver should be sized for the maximum BHP the pump can develop, since, as we will see, the centrifugal pump can vary its capacity and head, and, with these, the required BHP.

A centrifugal pump is normally designed for a specific flowrate (capacity) and head range. The pump casing can accommodate impellers of different diameters, again within a specific range.

A centrifugal pump can be driven at different speeds. With an impeller of a specific diameter and speed (RPM), the pump is capable of providing a range of capacities and heads (see 9.1).

For each flowrate, there is normally one corresponding head for the specific speed. Positive displacement pumps (piston, vane, etc.) will transfer a fixed capacity—the pump speed controls it—against any head, with properly sized driver.

Centrifugal pump manufacturers provide pump curves called Head-Capacity performance curves (H-C or H-Q curves). See chapter 10.

8.2. NET POSITIVE SUCTION HEAD (NPSH)

The capability of a centrifugal pump to transfer liquids within a specified range of capacities is predicated upon the energy available at its suction nozzle. The larger the capacity, the larger the energy required at the pump suction nozzle.

A centrifugal pump should be started only when its casing is full of liquid and devoid of air and vapors. When this condition is fulfilled, the pump is considered "primed".

The rotating impeller will transfer the liquid into the discharge pipe and generate very low pressure on its suction side. Because of the very low pressure (vacuum), new liquid will follow immediately. The liquid is brought to the pump by whatever energy is available at the suction nozzle.

The physical characteristics of the pump, the shape of its impeller, and its rotational speed will allow a specific range of capacities to flow through the pump. The precise capacity is governed by the Total Head for the specific system. In some cases, below atmospheric pressures are reigning at the suction nozzle. In order to avoid negative numbers to express heads or pressures, absolute units are used (PSIA or feet of liquid).

It is vital for the performance of the pump not to allow vapors in the suction line. Refer to chapter 1., item 14. This, as explained, will happen if the pressure in the suction line drops below the vapor pressure for the temperature at which the liquid is being pumped.

The pump manufacturers will indicate the condition at which they expect the pumped liquid to enter the pump's suction nozzle; in the first place, there should not be any vapors. This condition alone requires that the pressure at the pump nozzle be at least higher than the vapor pressure for the temperature of the liquid. On top of this requirement, as indicated above, a specific energy is required for each capacity.

Manufacturers indicate the required energy at the suction nozzle as a function of the pumped capacity. It is expressed in feet of water at 60° F. Water at 60° F will require more energy than other liquids. Hydrocarbons require a little bit less, but the designer is advised to use the "water" requirements in order to be on the conservative side. Some designers feel tempted to transform the indicated required suction head into feet of the liquid pumped (by dividing the indicated feet of water by the sp.gr. of the liquid). This would be too conservative for liquids lighter than water, and below requirements, if they are heavier than water.

The energy at the suction nozzle of the pump required to force the liquid into it is called NET POSITIVE SUCTION HEAD (NPSHR). Since the pump characteristic curves (see chapter 9) make allowance for the dissipation of this energy inside the pump, it must not be considered when computing the Total Head.

The available energy at the suction nozzle (or pump centerline), in conjunction with the condition of vaporless liquid, is called NET POSITIVE SUCTION HEAD AVAILABLE (NPSHA). Both NPSHR and NPSHA are measured in feet absolute. For a pump to operate properly, the NPSHA must be larger than the NPSHR.

From the above definitions:

NPSHA = Energy at pump suction (centerline) minus vapor pressure for the temperature at which the liquid is pumped.

The energy at the pump suction is computed by means of Bernoulli's equation.

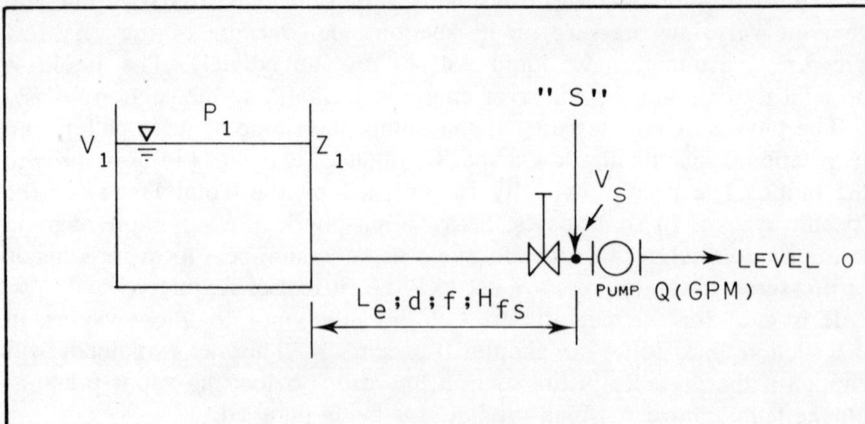

FIG. 8.01

Energy at Pump Suction and NPSHA:

$P_1+V_1^2/2g+Z_1-H_{fs}=P_s+V_s^2/2g+O=$energy at "S"

$$P_1+O+Z_1-f\frac{L_e}{d^5}\times\frac{Q^2}{32.18}-\text{Vapor pressure}=\text{NPSHA}$$

$$P_1+Z_1-f\frac{L_e}{d^5}\times\frac{Q^2}{32.18}-V_p=\text{NPSHA} \quad \text{8–03}$$

Note: Energy at $S=P_s+V_s^2/2g=P_1+Z_1-H_{fs}$

■ EXAMPLE 8:1

Refer to fig 8.01. Compute the energy, pressure, and NPSHA at point S for the following conditions:

$P_1=14.7$ psia, $Z_1=20$ FT of 80° water; sp.gr.$=1.00$
Vapor Pressure of 80° water$=0.5$ psia$=1.15$ FT
$L_e=100$ FT, ID$=6$ inches, f$=0.02$, steel pipe, Q$=400$ GPM; $V_s=4.54$ FPS; and $V_s^2/2g=0.32$ FT

$$\text{Energy at } S=P_1+Z_1-f\frac{L_e}{d^5}\times\frac{Q^2}{32.18}=P_s+V_s^2/2g$$

$$=14.7\times2.31+20-0.02\frac{100}{6^5}\times\frac{400^2}{32.18}$$

$=33.96+20-1.28=52.68$ FT ABS$=22.80$. PSIA Ans.

Pressure at $S=P_s=$Energy at $S-V_s^2/2g=52.68-0.32=52.36$ FT ABS.
$=22.67$ PSIA Ans.

Note: A gauge at S would indicate $22.67-14.70=7.97$ psig.
NPSHA$=$Energy at S$-$Vapor Pressure$=52.68-1.15$
$=51.53$ FT ABS. Ans.

■ EXAMPLE 8:2

Refer to Example 8:1. Compute the NPSHA for the following conditions at point S:

Temperature of water: 200° F, Specific gravity of 200° F water: 0.96, Vapor Pressure of 200° F water: 11.53 psia$=11.53\times2.31/0.96$ $=27.74$ FT ABS.

NPSHA$=P_1+Z_1-H_f-V_p=14.7\times2.31/0.96+20-1.28-27.74$
$=26.35$ FT. Ans.

In case the NPSHR of a specific pump exceeds 26.35 FT, the pump will fail to transfer 400 GPM. In general, when the NPSHR is larger than the NPSHA, the pump will operate unstably. It will cavitate.

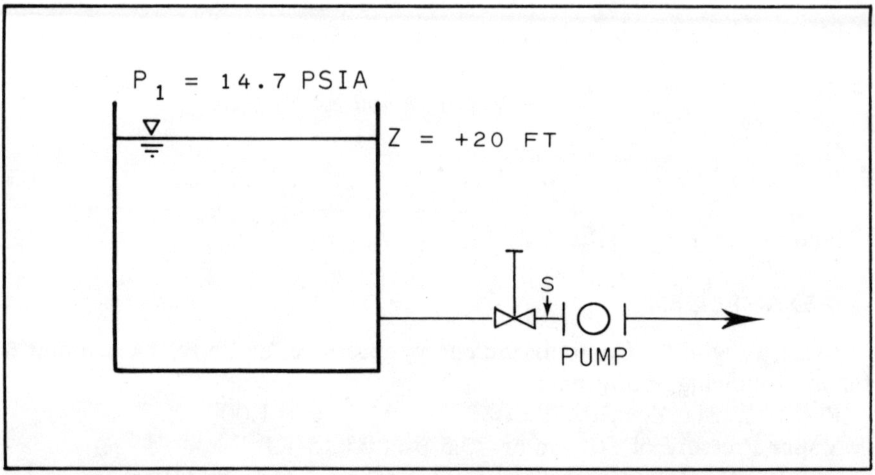

FIG. 8.02

■ EXAMPLE 8:3

Compute NPSHA per fig. 8.02 and conditions indicated: water at 212° F, sp.g. = 0.95, vapor pressure = 14.7 PSIA, $L_e = 100$ FT, ID = 6 IN.; $h_f = 1.28$ FT, Q = 400 GPM, and $Z_1 = +20$ FT.

NPSHA = Energy at S − Vapor Pressure

$= 14.7 \times 2.31/0.95 + 20 - 1.28 - 14.7 \times 2.31/0.95$

$= 18.72$ FT ABS Ans.

In case the NPSHR exceeds 18.72 FT, the pump will cavitate. In order to avoid cavitation, Z_1 has to be increased. If the NPSHA per above is satisfactory for Z = +20 FT, it may not be when the water level will drop below a critical value. Level control is then mandatory.

■ EXAMPLE 8:4

Compute NPSHA per arrangement on fig 8.03 and indicated conditions: Q = 400 GPM; ID = 6 in; $h_f = 1.28$ FT; $L_e = 100$ FT; water at 80° F (sp.g. = 1.0), vapor press.: 0.5 psia = 1.15 FT, and Z = −10 FT.

NPSHA = Energy at S − Vapor pressure $= P_1 + Z_1 - H_f - V_p$

$= 14.7 \times 2.31 - 10 - 1.28 - 1.15 = 33.96 - 12.43 = 21.53$ FT ABS Ans.

The NPSHR of the pump has to exceed 21.53 FT at all times.

■ EXAMPLE 8:5

For an arrangement per fig 8.03 and conditions per example 8:4, compute the maximum water temperature at which a flowrate of 400 GPM can take place without cavitation (Pump NPSHR = 15 FT).

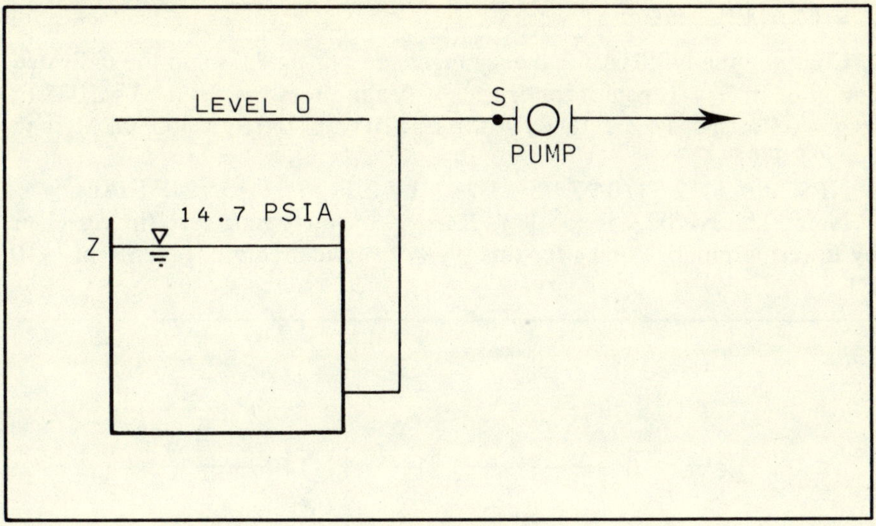

FIG. 8.03

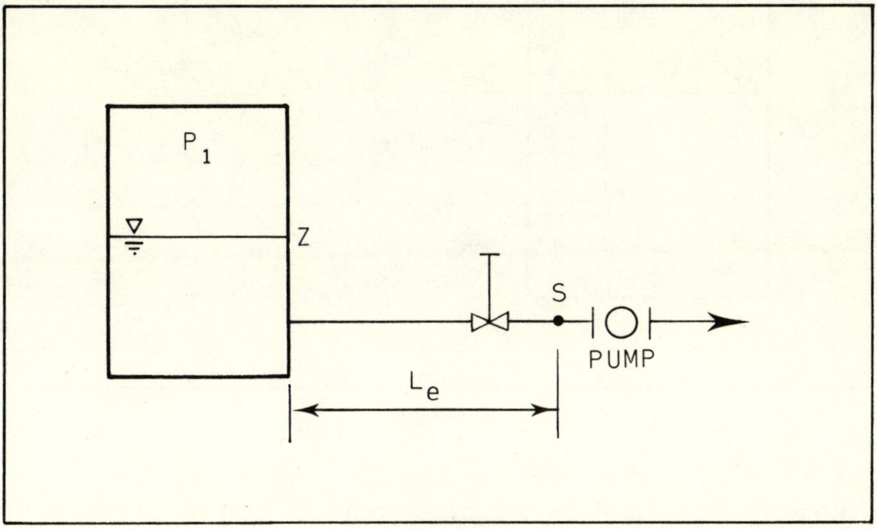

FIG. 8.04

NPSHA = Energy at S − Vapor Pressure = $14.7 \times 2.31 - 10 - 1.28 - V_p$
$= 22.68 - V_p = 15$
$V_p = 7.68$ FEET = 3.32 PSIA
(**Note:** The difference in sp.gr. of water has been neglected).
From steam tables: $T_{MAX} = 146°$ F. (Boiling Temp. for 3.32 PSIA) Ans.

■ EXAMPLE 8:6

Compute the NPSHA for the arrangement per fig 8.04 and the indicated conditions: $P_1=1$ psia (condensate), Vap. press.$=1$ psia, $T=102°$ F; sp.g.$=0.97$; $h_f=1.28$ FT; $L_e=100$ FT; ID$=6$ IN; $Q=400$ GPM, and $Z=+5$ FT.

NPSHA$=1\times2.31/0.97+5-1.28-1\times2.31/0.97=3.72$ FT Ans.

Note: The NPSHA is too low. Raise Z by increasing level in vessel or by lowering pump. Pumps for this type of service require a NPSH of 6-10 FT.

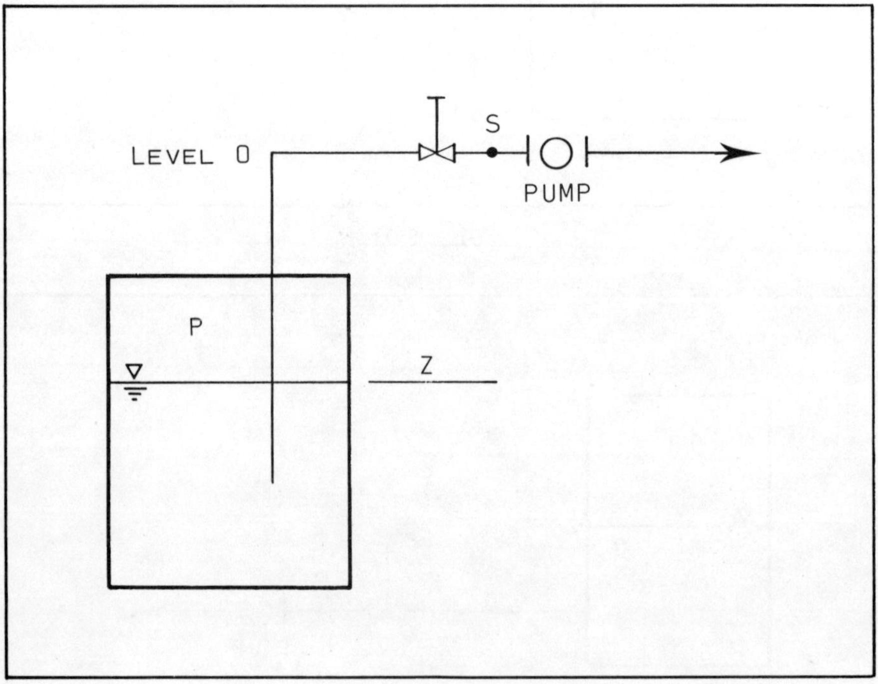

FIG. 8.05

■ EXAMPLE 8:7

Compute the NPSHA for the arrangement per fig. 8.05 and the indicated conditions: Liquid$=$Butane, Temp$=90°$ F, $P=60$ psig, sp.gr.$=0.6$, $V_p=44$ psia (90° F), $H_f=15$ FT, and $Z=-10$ FT.

NPSHA$=$Energy at $S-V_p=60\times2.31/0.6-10-15-44\times2.31/0.6$

$=206-169.4=36.6$ FT ABS Ans.

Note: The NPSHR of the pump has to be lower than 36.6 FT. of water.

9.

Characteristic Pump Curves

Centrifugal pump manufacturers use composite pump characteristic curves (graphs) to represent the average performance of a specific type pump at a fixed speed (RPM). The curves are obtained by means of testing. (Head-capacity Curves).

The following four parameters are of major importance to the application engineer for the proper selection and application of a suitable pump. (Refer to fig. 9.01).

9.1. TOTAL HEAD AS FUNCTION OF THE CAPACITY (H-C CURVE)

It should be borne in mind that the curves provided by manufacturers are the NET pump H-C (Head-Capacity) curves, i.e., they do not indicate the internal pump losses which are precalculated and indicated as efficiency curves.

The losses in a pump consist of: pure mechanical, internal leakage, so called skin friction, shock losses at the entrance in the impeller, and other disturbances. As indicated previously, the losses are taken care of by the driver.

As far as the application engineer is concerned, the H-C curve he sees, represents the ability of the pump to do external work, i.e., provide the energy to transfer liquids from one point to another. H-C curves shift with the impeller diameter. The unit selected for the capacity Q is GPM, and for the head H is FEET.

FEET OF LIQUID PUMPED. The head represents the work per mass unit of liquid pumped (in feet). This unit is very practical because the H-C curve, for a specific impeller diameter and fixed RPM, will not shift, regardless of the characteristics and type of liquid pumped. The relationship between capacity and the corresponding head on the H-C curve will remain constant. The head is defined as Total Head, or Differential head,

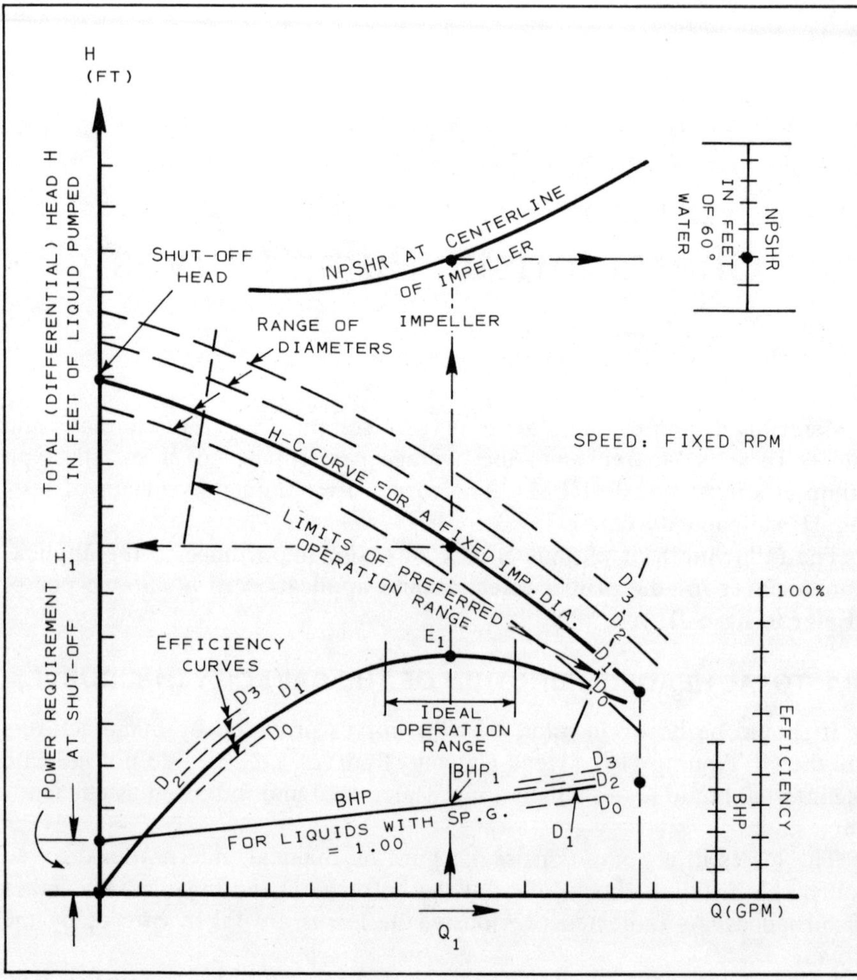

FIG. 9.01
Characteristic pump curves.

expressed in feet of liquid pumped. Manufacturers do not add the crucial words "of liquid pumped" which may induce the assumption that the head is in feet of water which, if used as basis, would distort the expected performance.

The Total Head indicated on pump H-C curves represents the value of E_p, as computed in previous chapters.

Developing the same head in feet for any liquid doesn't obviously mean that the specific head, expressed in PSIA, would nearly be the same. A pump developing a head of 100 FT will produce different pressures, depending on the specific gravity of the liquid. From a theoretical point of view, we could choose to represent the H-C curves in GPM versus PSIA. In this case, one has to construct a curve for each specific gravity to be considered, a difficult and impractical task.

Each H-C curve is plotted for a specific impeller diameter and rotational speed (RPM). In most curves, for each Q, there is only one H value on the pump curve. There are, though, pump curves which droop. In this case, the highest point is not at Q=O (shut-off), and there can be two Q values for a certain H. It should be stressed that the preferred curve shape is the one for which there is just one Q for each H; in other words, a steady increase of H with decreasing Q.

The stable operating range for a pump extends from a minimum capacity (10-15% of the capacity at maximum efficiency) to a maximum capacity (approximately 5% less than indicated on the H-C curve).

Operating a centrifugal pump with a closed discharge valve is allowed, but for only very short intervals. The capacity output is zero and, in most cases, the head is the maximum (shut-off). At shut-off, a liquid will tend to boil off which would be detrimental for the pump.

Operating the pump below minimum capacity per above could also result in heat-up and vaporization.

As mentioned before, the head-capacity curve is obtained by testing. With increasing capacity, the head which the pump can develop decreases, so equation 8–02 can be satisfied for each point.

The reader should use formula 8–02 and compute values off a sample pump curve in chapter 10. Analyzing the H-C curve on Fig. 9.02, we can say that the pump can deliver up to Q_1 GPM, and generate heads from H_1 to H_s. The lesser the Q, the higher the head which can be made available and vice-versa. In order for the pump to transfer Q_1 GPM, the total head should not exceed H_1. In case it does, (H_2), the pump will only deliver Q_2, which is less than Q_1. With still higher total head (H_3) Q will continue to decrease (Q_3). The head is the highest for Q=O (H_s), (shut-off head). The shut-off head doesn't anymore represent the total system head, because there is no flow. For detailed analyses of pump curves, refer to chapter 10.

Operating a pump at maximum Q (extreme point on curve) results in higher NPSH required by the pump. If not available, the pump will cavitate, and pitting may occur. This condition is very detrimental for the pump and should be avoided at all cost. Ideally, a pump should operate within a narrow range of the area of its maximum efficiency.

The shape of the pump curve (H-C) is indicative to the application engineer as to what to expect in case of deviations and miscalculations. Also, it is vital for the prediction of the pump operation when changes in the system are introduced.

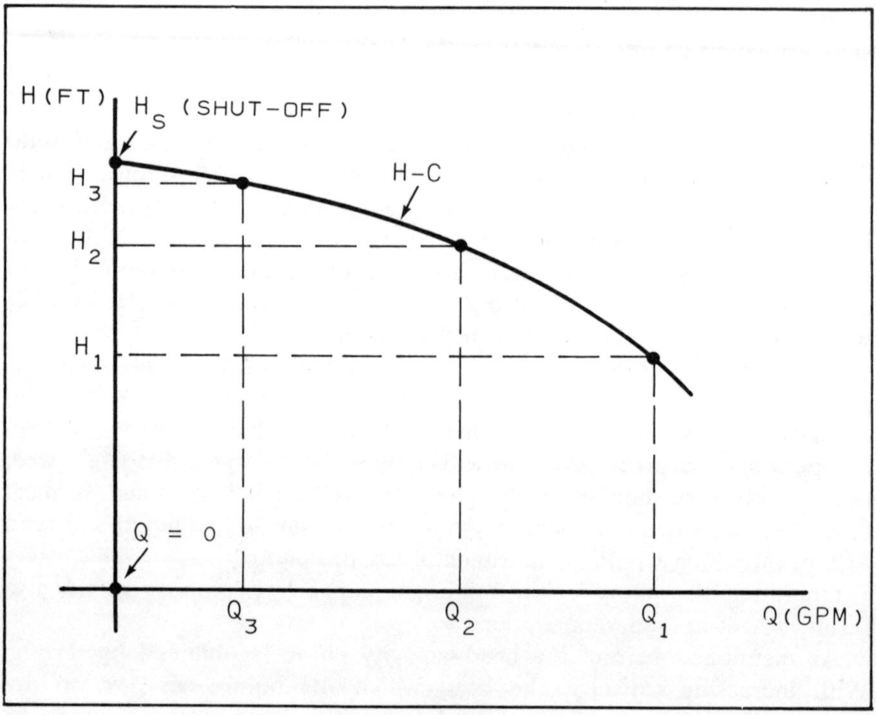

FIG. 9.02
Head-capacity curve.

It must be said that, although a pump curve is easy to understand and analyze, many pumps have been bought and installed without proper analysis of the H-C curve. The result: poor performance and frequent replacements of parts, and of the pump itself.

9.2. BHP CURVES

The BHP curves indicate the BHP required by the pump at the pump shaft. The BHP represents the horsepower required to provide the energy to transfer liquids from one point to another, and the energy required to overcome the pump losses. It can be calculated for each point of the H-C curve, considering the efficiency at it, by means of formula 8–02:

BHP = H × Q × Sp. gr./(3,960 × efficiency)

where head is in feet of liquid, capacity is in GPM, and efficiency is in percent (fraction).

Per example 7:1, using formula 8–02, for Q = 2,000 GPM, H = 310 FT of liquid and an efficiency of 75%:

BHP = 2,000 × 310/(3,960×.75) = 208.75, as calculated before.

It should be noted that the BHP curves are plotted for liquids with a specific gravity of 1.00. For a different specific gravity, the value on the curve should be multiplied by it.

When figuring the horsepower requirements of the driver, one should always assume that the pump may operate at one time or another at a point for which the BHP is the highest. The maximum BHP should be taken into consideration in order not to overload the driver. This is especially important for electrical motors.

The BHP at zero flow (Shut-off) is not zero. This is because energy is expended to rotate the impeller(s) in the stationary liquid. Some pump types (propeller, mixed flow) have their highest BHP requirements at shut-offs. In this case, the operation at shut-off is not advisable.

9.3. EFFICIENCY CURVES

Expressed in percentage, the efficiency for each point on the H-C curve represents the ratio of the energy imparted by the pump to the liquid to the energy obtained by the pump at the pumpshaft. The efficiency is also an indicator of the energy lost in the pump. The higher the efficiency, the lower the losses.

9.4. NET POSITIVE SUCTION HEAD REQUIRED (NPSHR)

The NPSHR is indicated in feet of water at 60°F. As explained, pumping water at 60°F by a particular pump would require more NPSH than other liquids. In order to be on the conservative side, one should use the "water" NPSH. One should not translate feet of water into feet of pumped liquid. For viscous liquids, see Chapter 11.4.

Some manufacturers indicate that the NPSHR is referred to the centerline of the impeller. The application engineer can compute the NPSHA at the pump suction nozzle, but not at the centerline of the impeller. Therefore a foot or two should be added to the indicated NPSHR value.

When designing a pumping system in which a pump will be taking suction from a storage tank or vessel, the computation of the NPSHA should allow for minimum level in the tank, about 3 FT above its suction nozzle. This would permit the pump to operate with abundant NPSHA all the time.

9.5. Although not indicated on pump curves, an important relationship exists between the impeller diameter, the head developed, and the speed of the pump (RPM):

$$D = 1{,}840 \times H^{0.5}/RPM \quad \text{.......... 9–01}$$

where:

D = impeller diameter, in inches

H = head developed by pump, in feet of liquid pumped,

RPM = pump speed

■ EXAMPLE 9:1

With 3,500 RPM, the discharge pressure of a pump indicated by the gauge is 160 psig. The suction pressure, also indicated by the gauge, amounts to 5 psig. Neglecting the differential between the velocity heads on the suction and discharge sides, establish the diameter of the pump in operation.

The head: = 160–5 = 155 PSI = 155 × 2.31 = 358 FT of water

$D = 1{,}840 \times 358^{0.5}/3500 = 9.94$, say 10 inches. Ans.

Note: Formula 9–01 is approximate. It is preciser if the operating point is within the range of high efficiencies. Refer to figure 10.11: for a head of 358 feet, the pump will deliver 950 GPM with an impeller of 10.25 in.

10.

Evaluation of Manufacturers' Pump Performance Curves

Twenty-four centrifugal pump characteristic curve groups from eight manufacturers, selected at random, are shown in figs. 10.01 through 10.24.

The reader is advised to study these pump curves very carefully in conjunction with the tabulated evaluation summaries in tables 10.25, 10.26, and 10.27 which represent a detailed analysis of the information supplied on these pump-curve sheets.

This is the type of analysis the application engineer is expected to perform before selecting a pump for a specific system, especially when comparing different pumps offered for the same service. In addition to this, the application engineer should study the particular pump catalogs for technical details and discuss the specific pumps with the manufacturer. The following tabulation lists the manufacturers whose pump characteristic curves are included in this chapter:

Manufacturer	Figure Numbers
1. Weinman Pump—LFE Corp. FCD	10.01, 10.02, 10.08, 10.09
2. Goulds Pump, Inc., Seneca Falls, NY. 13148	10.03, 10.04, 10.05, 10.06, 10.16, 10.17
3. Pacific Pumping Co., Oakland, Ca.	10.07, 10.10
4. Aurora Pump, A Unit of General Signal Corp.	10.11, 10.12, 10.24
5. Crane Deming Pumps	10.13, 10.14
6. Afton Pumps Inc.	10.15
7. Johnston Pump Co.	10.18, 10.23
8. United Centrifugal Pumps	10.19, 10.20, 10.21, 10.22

1750 R.P.M.

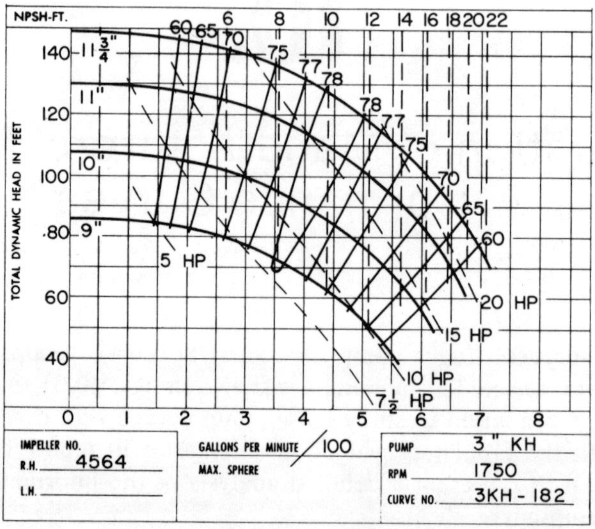

FIG. 10.01

Courtesy: Weinman Pump—LFE Corp. FCD

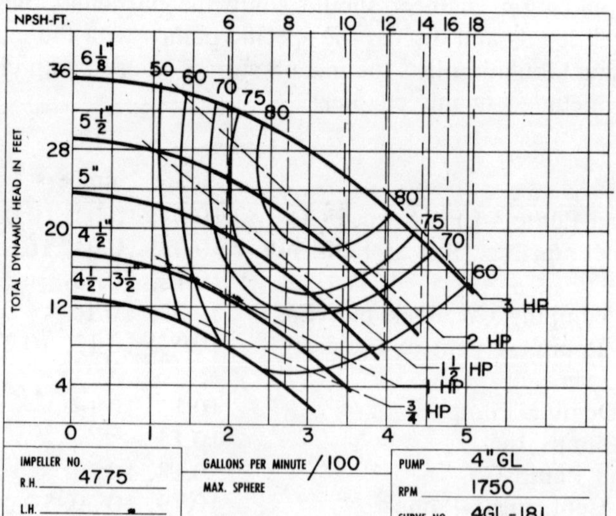

FIG. 10.02

Courtesy: Weinman Pump—LFE Corp. FCD

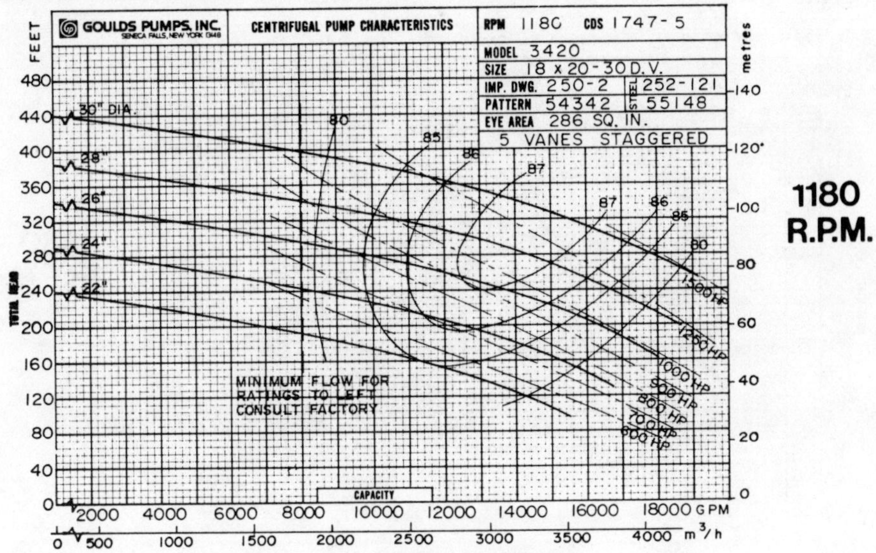

FIG. 10.03

Courtesy: Goulds Pumps Inc., Seneca Falls, N.Y. 13148

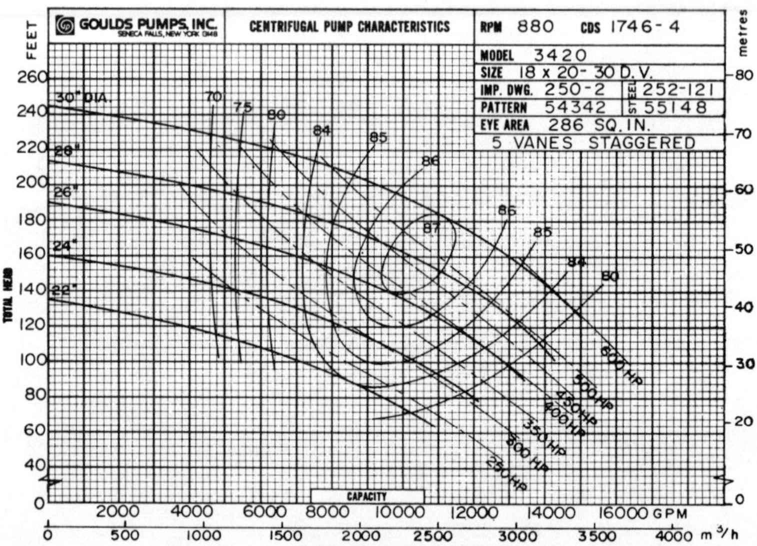

880 R.P.M.

FIG. 10.04
Courtesy: Goulds Pumps Inc., Seneca Falls, N.Y. 13148

1750 R.P.M.

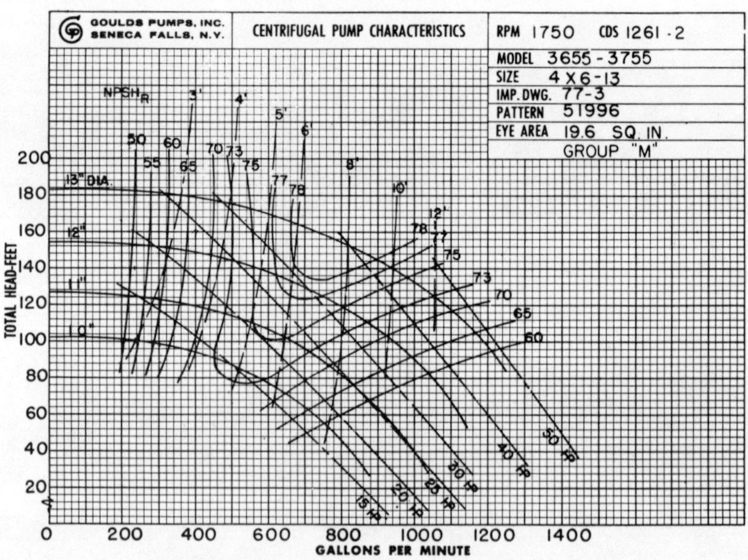

FIG. 10.05
Courtesy: Goulds Pumps Inc., Seneca Falls, N.Y. 13148

1150 R.P.M.

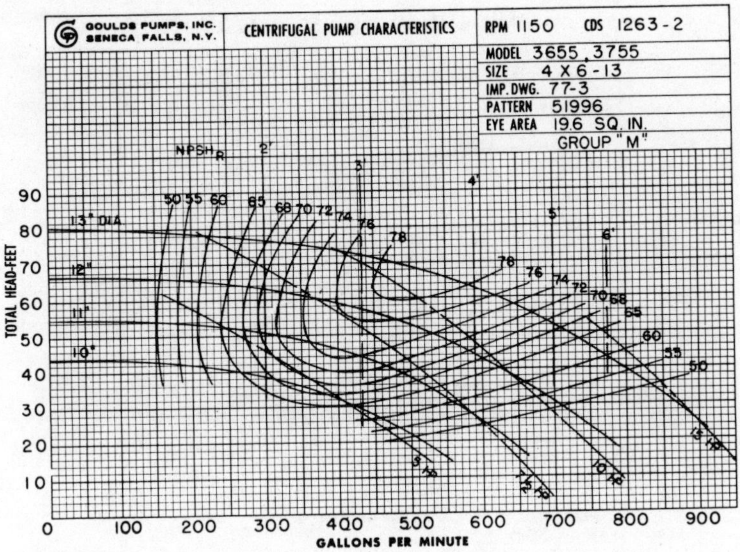

FIG. 10.06

Courtesy: Goulds Pumps Inc., Seneca Falls, N.Y. 13148

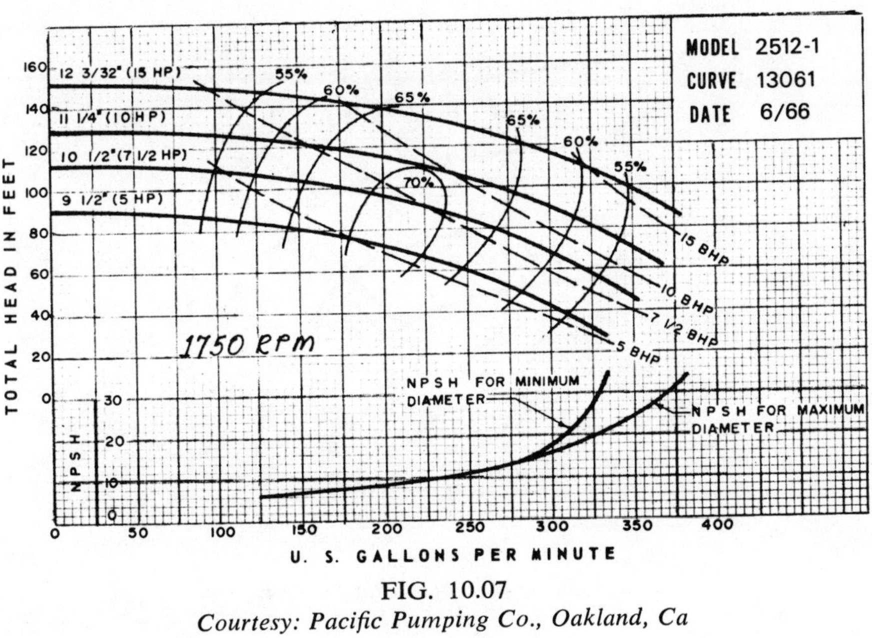

FIG. 10.07

Courtesy: Pacific Pumping Co., Oakland, Ca

OPEN IMPELLER – 1750 R.P.M.

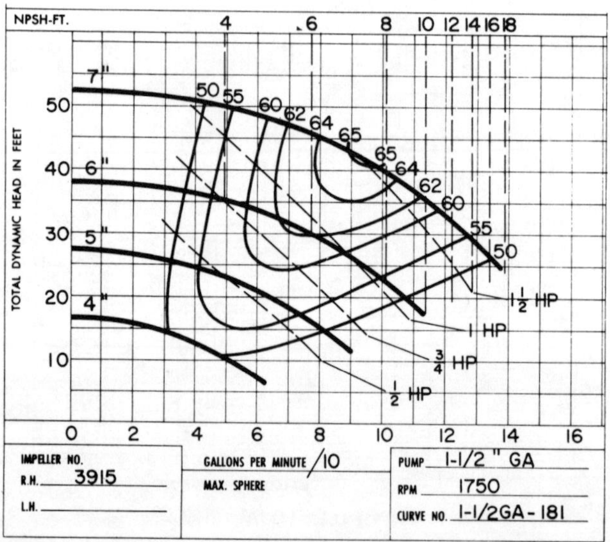

FIG. 10.08
Courtesy: Weinman Pump—LFE Corp. FCD

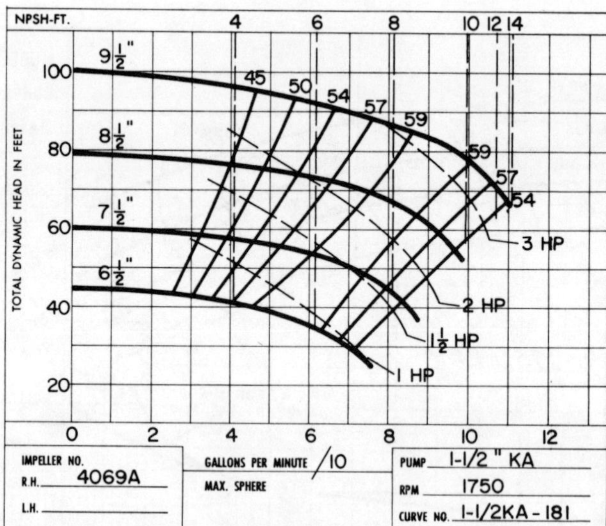

FIG. 10.09
Courtesy: Weinman Pump—LFE Corp. FCD

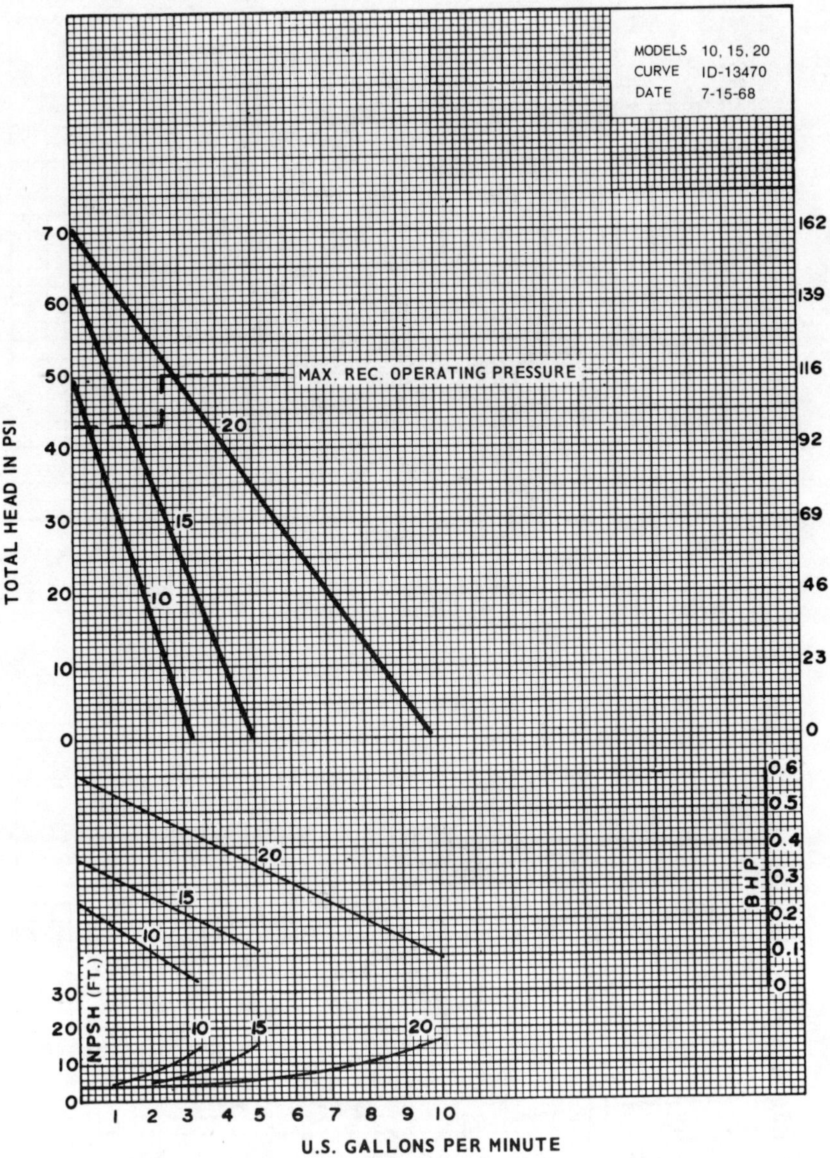

FIG. 10.10
Courtesy: Pacific Pumping Co., Oakland, Ca

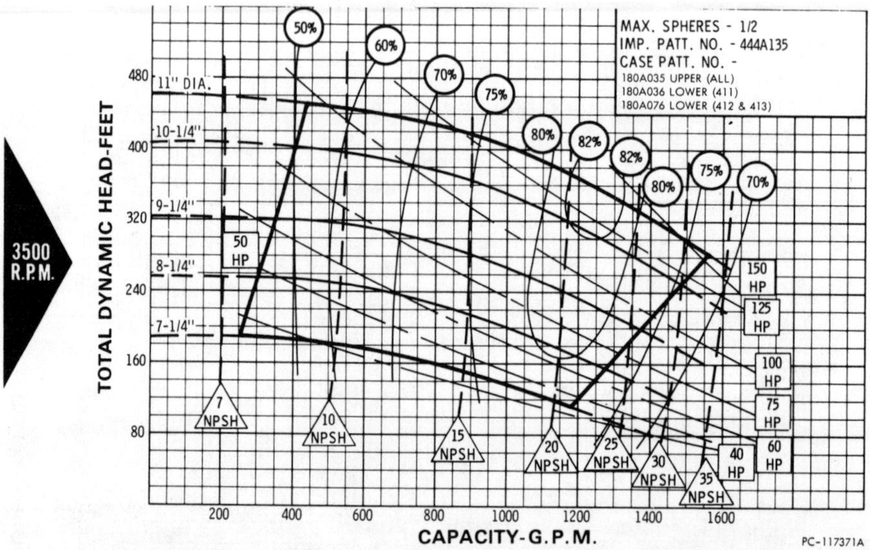

FIG. 10.11

Courtesy: Aurora Pump—A Unit of General Signal Corp.

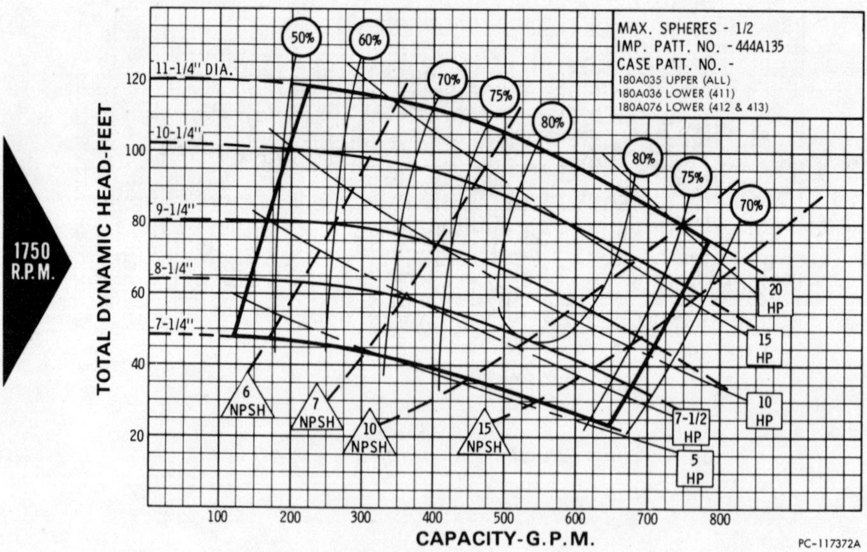

FIG. 10.12

Courtesy: Aurora Pump—A Unit of General Signal Corp.

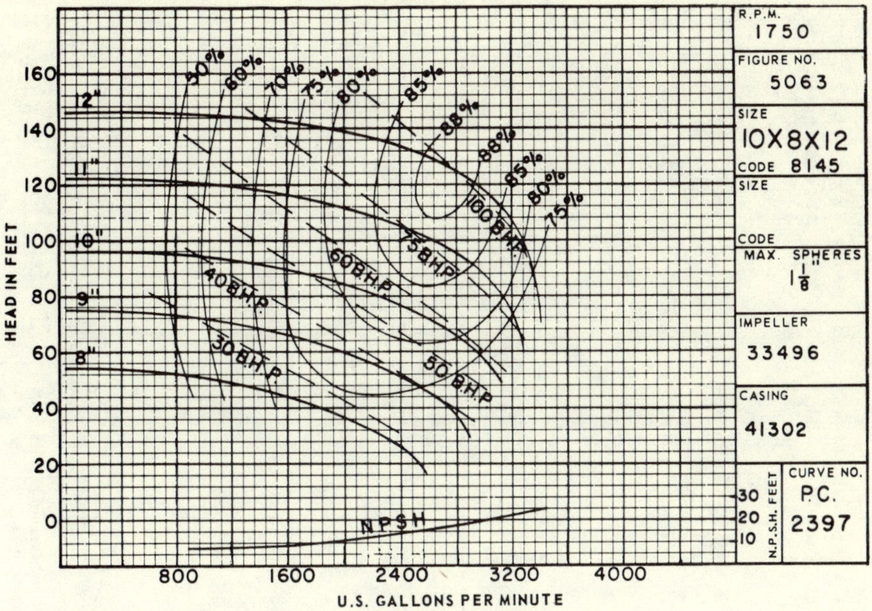

FIG. 10.13
Courtesy: Crane Deming Pumps

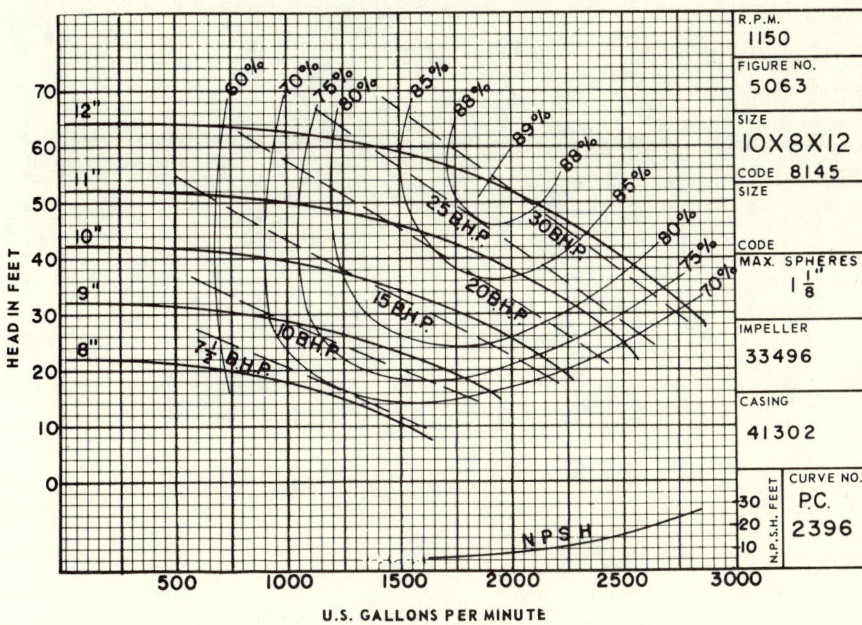

FIG. 10.14
Courtesy: Crane Deming Pumps

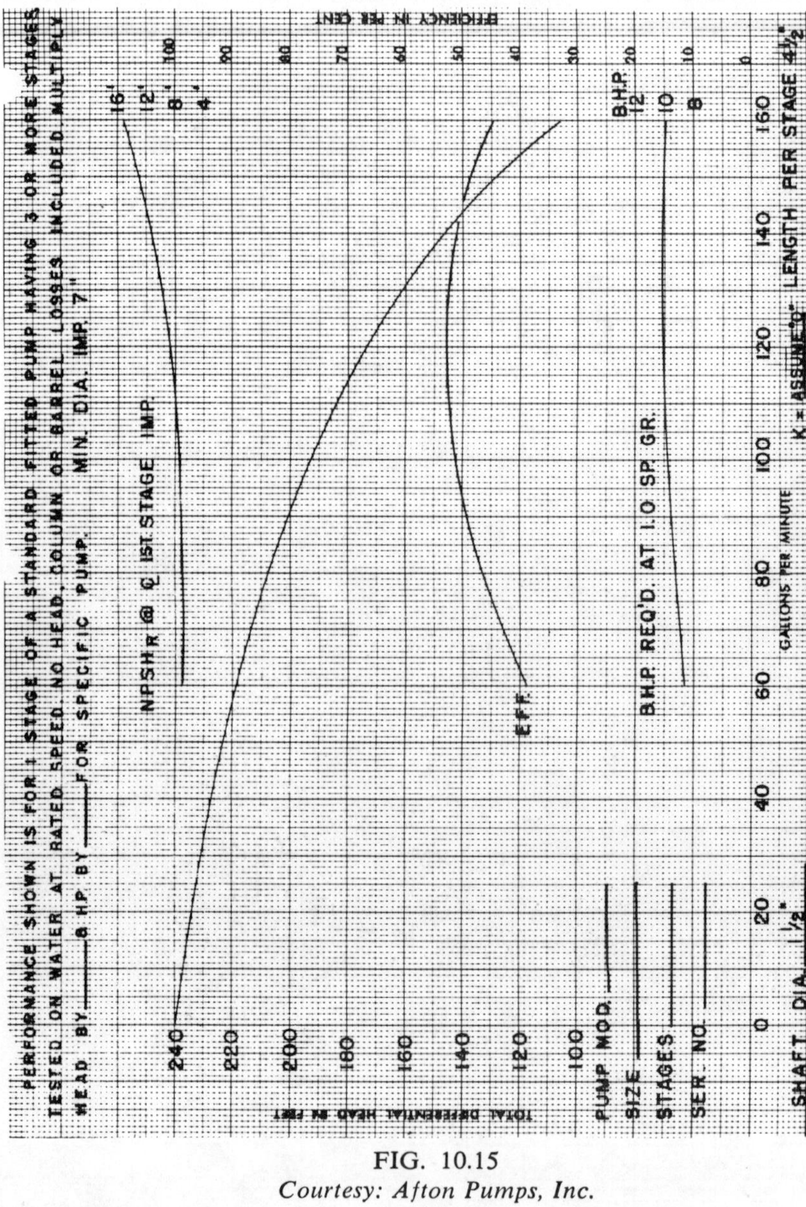

FIG. 10.15

Courtesy: Afton Pumps, Inc.

3550 R.P.M.

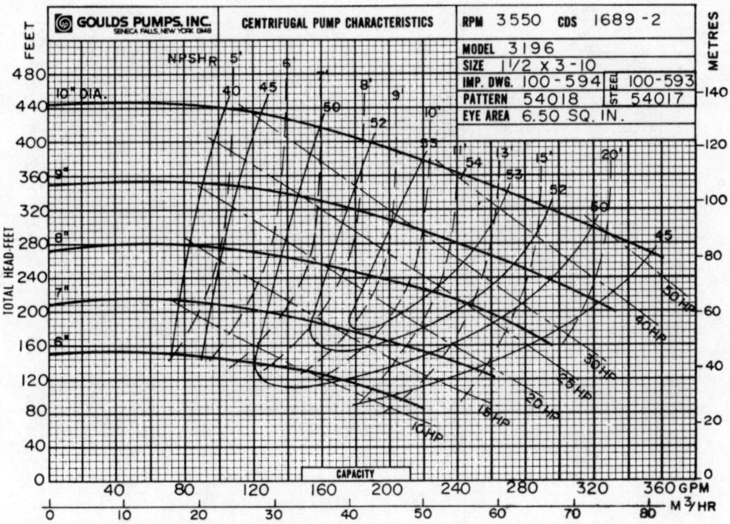

FIG. 10.16
Courtesy: Goulds Pumps Inc., Seneca Falls, N.Y. 13148

1750 R.P.M.

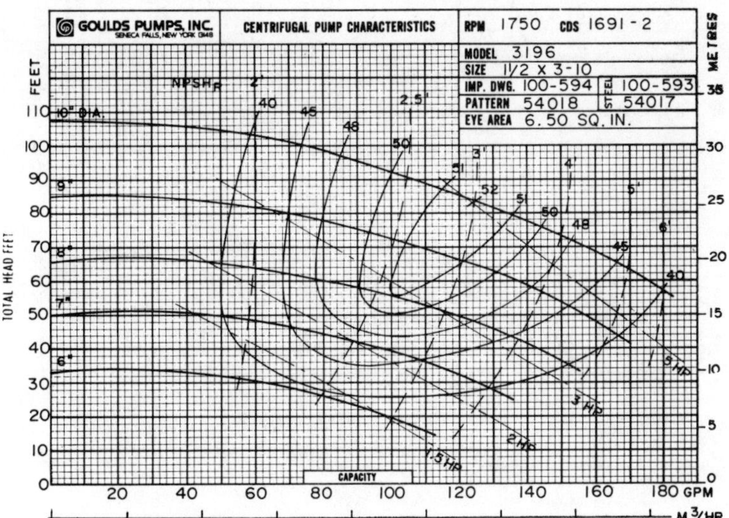

FIG. 10.17
Courtesy: Goulds Pumps Inc., Seneca Falls, N.Y. 13148

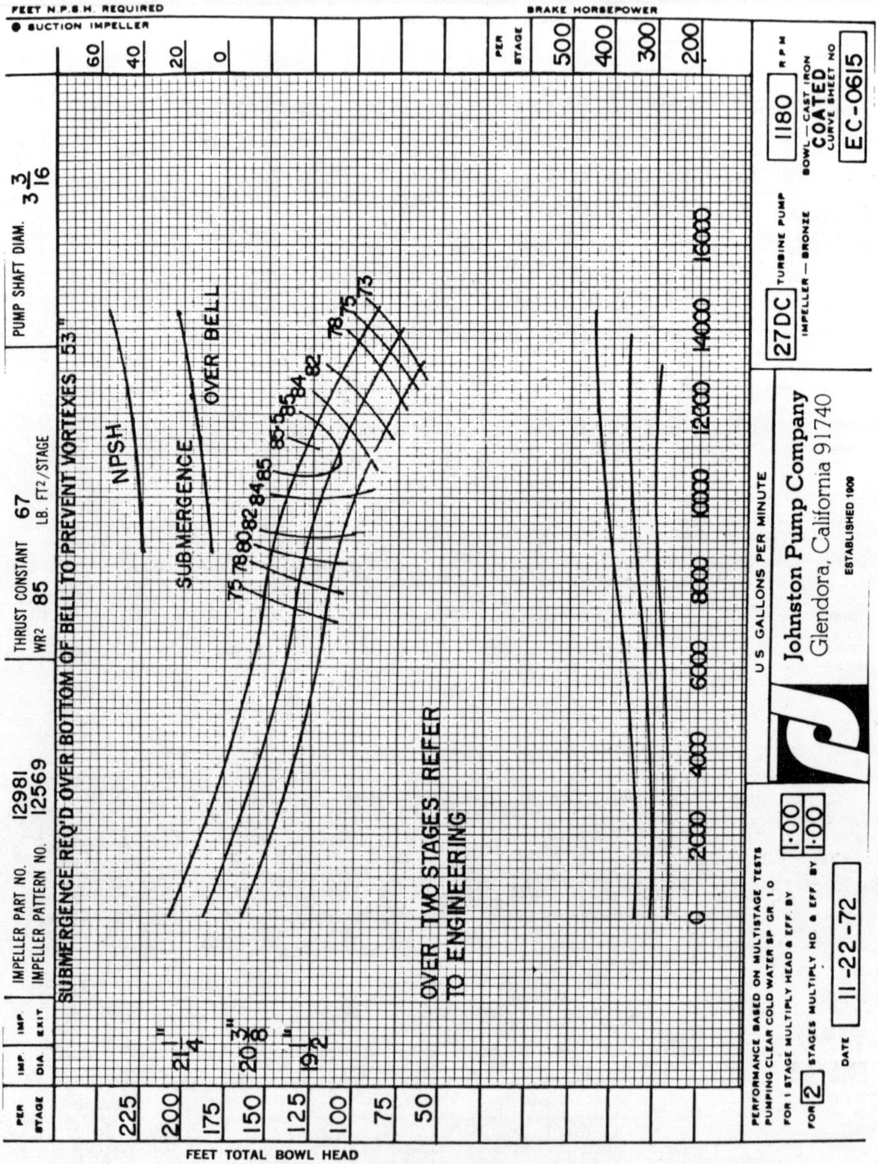

FIG. 10.18
Courtesy: Johnston Pump Company

1000-43

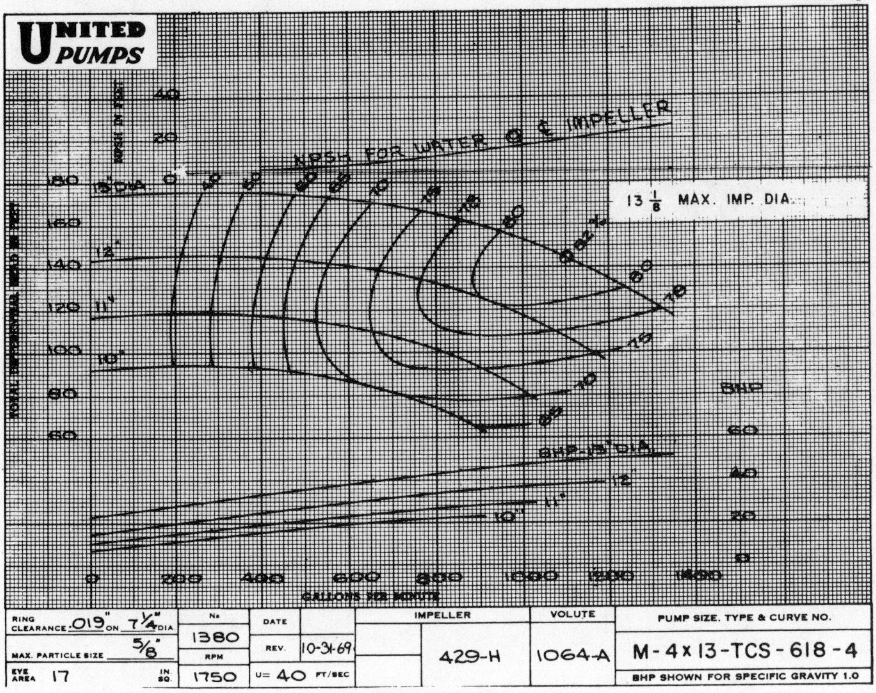

FIG. 10.19

Courtesy: United Centrifugal Pumps

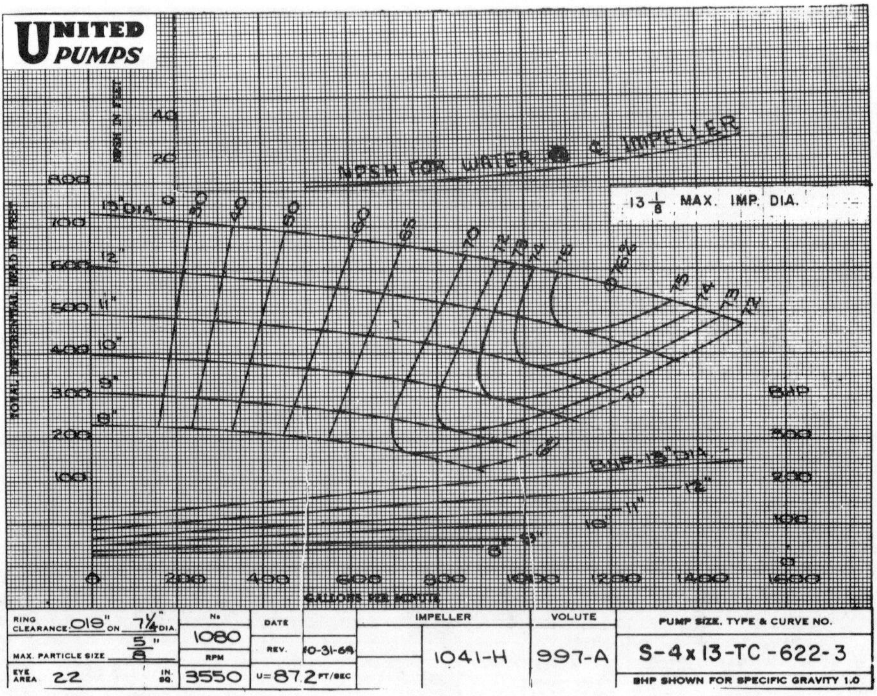

FIG. 10.20
Courtesy: United Centrifugal Pumps

NEW ISSUE JUNE 28, 1972 2800-47

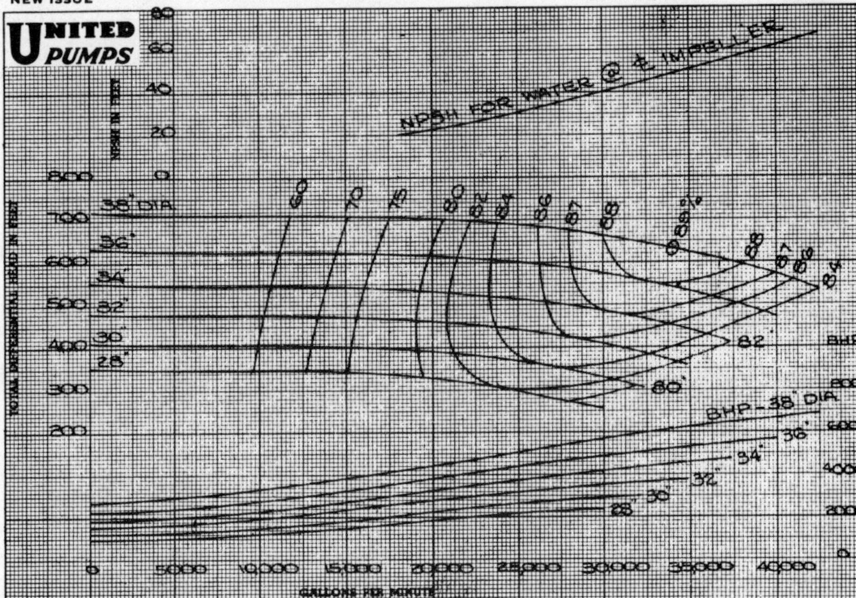

RING CLEARANCE .025" ON 20-1/8" DIA		Ns	DATE		IMPELLER	VOLUTE	CASE PATT.	PUMP SIZE, TYPE & CURVE NO.
		1738	REV.	6-28-72	1717-H	1842-A	2800-A 2800-A1	M - 24 x 38 DVS - 1278-2
MAX. PARTICLE SIZE 2-1/8"		WET WR² LBS. IN²						
© EYE AREA 390 IN. SQ.	u= 90.6 FT/SEC	182,500	RPM	1185				BHP SHOWN FOR SPECIFIC GRAVITY 1.0

FIG. 10.21
Courtesy: United Centrifugal Pumps

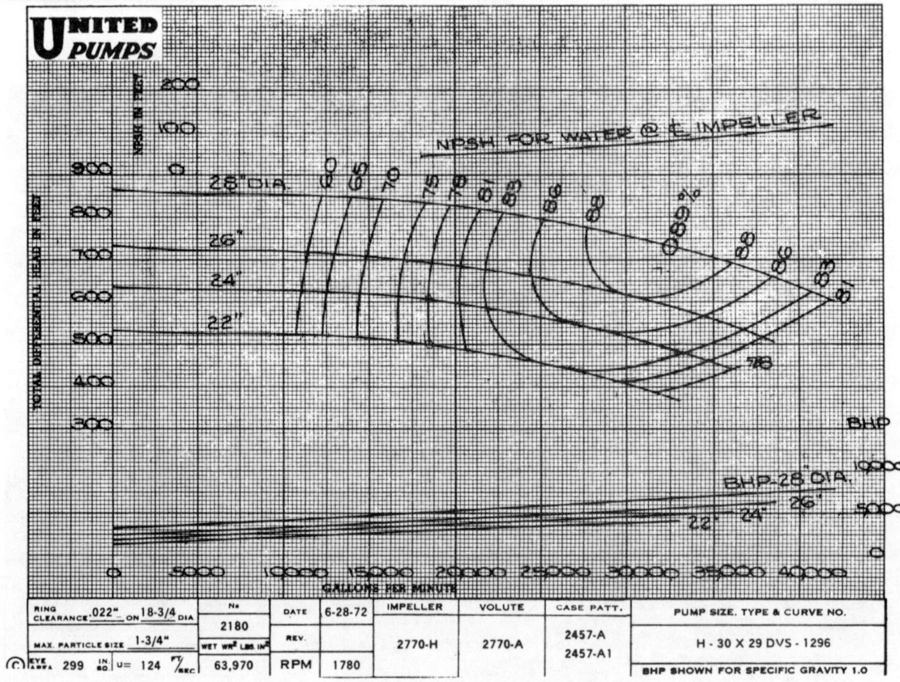

FIG. 10.22
Courtesy: United Centrifugal Pumps

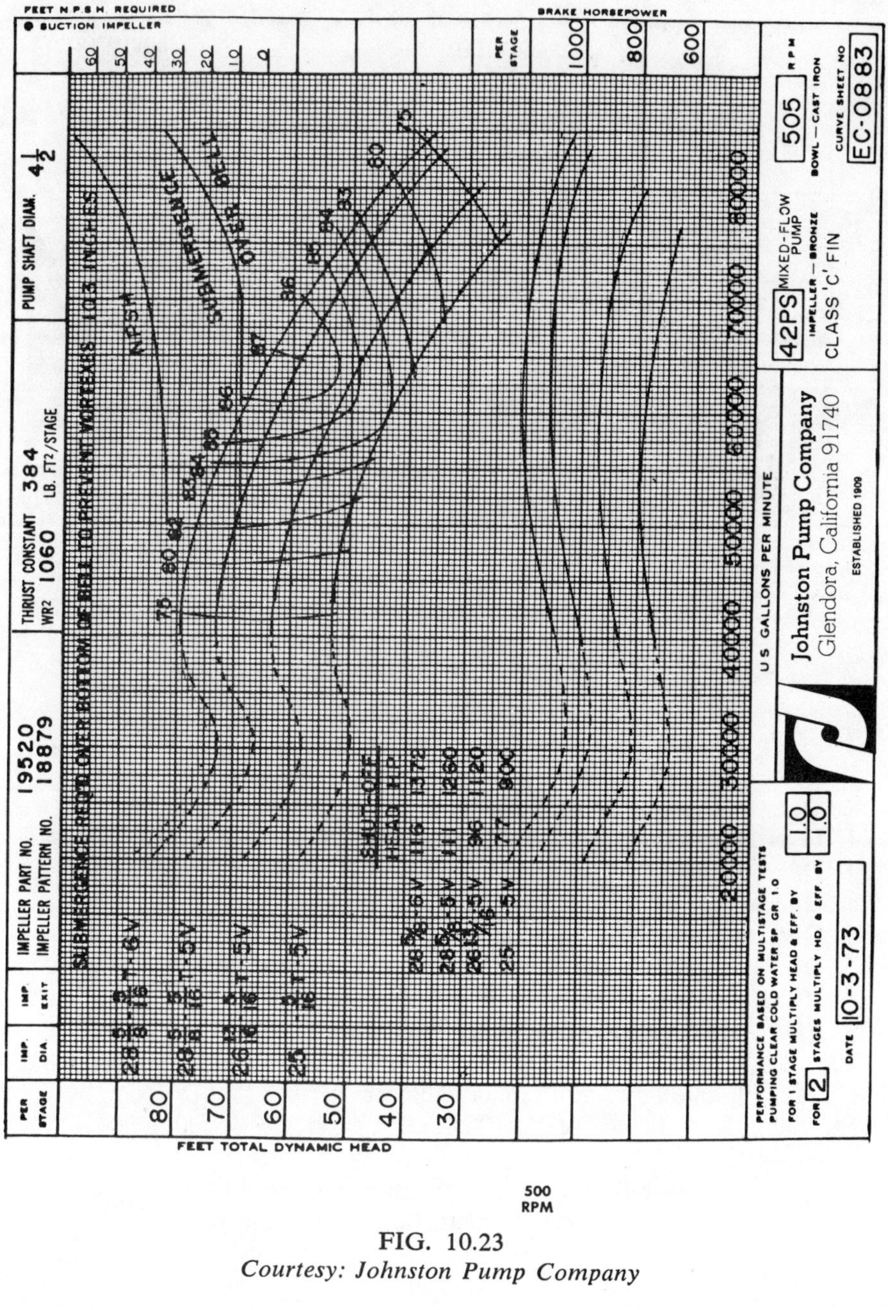

500
RPM

FIG. 10.23
Courtesy: Johnston Pump Company

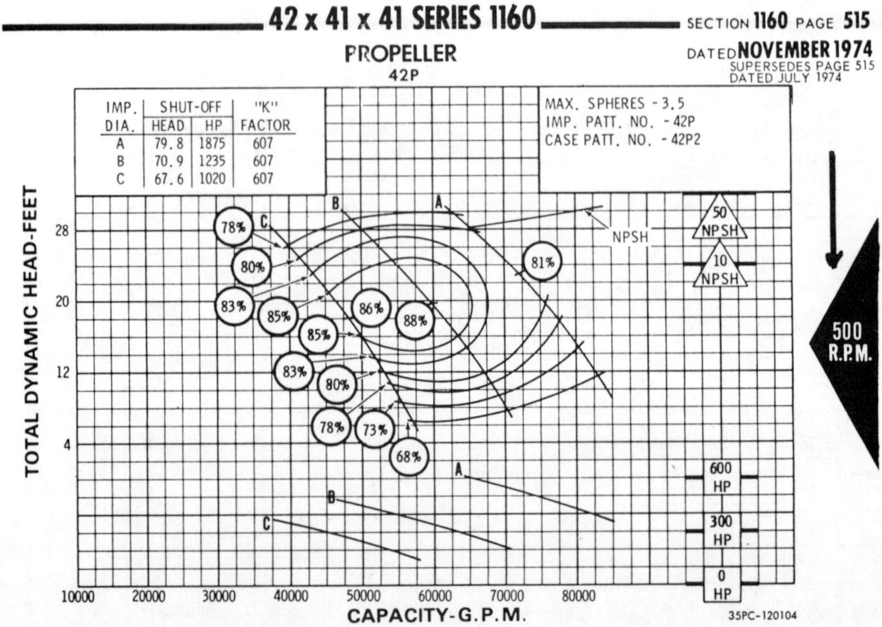

FIG. 10.24

Courtesy: Aurora Pump—A Unit of General Signal Corp.

The following abbreviations are used in Tables 10.25, 10.26, and 10.27:

1. D = Range of impeller diameter sizes, in inches.
2. Q_{max} = Range of maximum capacities achievable, in GPM.
3. H = Range of heads achievable, in feet of liquid pumped.
4. Q_E = Capacity range within range of higher efficiencies.
5. E = Range of efficiencies, in percent.
6. $Q_{E\text{-}M}$ = Capacity at maximum efficiency, in GPM.
7. $H_{M\text{-}EM}$ = Maximum head at maximum efficiency, in feet of liquid.
8. RPM = Pump speed (revolutions per minute).
9. BHP = Range of brake horsepower requirements.
10. $Q_{min/max}$ = Limits of recommended flowrate ranges (estimated). For exact limitations, consult manufacturer.
11. NPSHR = Range of Net Positive Suction Heads required, in feet of water.
12. NPSHR/D=NPSHR varies with impeller diameter (yes; slightly; no)
13. H_{SH} = Range of shut-off heads, in feet of liquid pumped.

TABLE 10.25
EVALUATION OF PUMP PERFORMANCE CHARACTERISTICS

	ITEMS	FIG. 10.01	FIG. 10.02	FIG. 10.03	FIG. 10.04	FIG. 10.05	FIG. 10.06	FIG. 10.07	FIG. 10.08
1	D	9-11¾	3½-6⅛	22-30	22-30	10-13	10-13	9½-12²³⁄₃₂	4-7
2	Q_{max}	550-700	300-500	15,000-20,000	10,000,15,000	850-1,200	500-850	325-375	60-130
3	H	45-145	4-33	*120-400	80-230	40-180	20-78	30-150	8-50
4	Q_E	350-550	200-450	10,000-17,000	7,000-14,000	500-1,100	400-700	150-300	60-110
5	E	60-78	50-87	80-87	70-87	50-78	50-78	55-70	50-65
6	Q_{E-M}	475	325	15,000	11,000	900	575	225	90
7	H_{M-EM}	125	27	335	182	165	70	110	45
8	RPM	1,750	1,750	1,180	880	1,750	1,150	1,750	1,750
9	BHP	5-25	½-3	550-1,500	225-600	15-50	5-15	5-17	⅜-2
10	$Q_{min/max}$	150/650	120-450	*8,000-17,000	5,000-13,000	400-1,100	250-800	125-325	45-120
11	NPSHR	4-22	4-18	Consult Mfr.	Consult Mfr.	3-14	2-7	6-33	4-18
12	NPSHR/D	No	No	Consult Mfr.	Consult Mfr.	Slightly	No	Yes	No
13	H_{SH}	85-150	13-25	240-440	136-244	104-185	44-81	90-150	17-53

*Indicated.

TABLE 10.26
EVALUATION OF PUMP PERFORMANCE CHARACTERISTICS

	ITEMS	FIG. 10.09	FIG. 10.10	FIG. 10.11	FIG. 10.12	FIG. 10.13	FIG. 10.14	FIG. 10.15	FIG. 10.16
1	D	6½-9½	Constant*†	7¼-11	7¼-11¼	8-12	8-12	7½ (min 7)	6-10
2	Q_{max}	70-110	3-10	‡1300-1500	‡650-840	2,500-3,300	1,500-2,800	160	220-360
3	H	30-95	‡10-100	‡80-440	‡25-110	20-142	10-63	¶110-230	90-440
4	Q_E	50-100	2-9	800-1,500	320-750	1,600-3,200	1,200-2,600	80-150	160-320
5	E	45-59	§10-30	50-82	50-80	50-88	60-89	40-53	40-54
6	Q_{E-M}	100	—	1,300	650	2,800	2,000	120	240
7	H_{M-EM}	85	—	370	100	130	54	¶174	360
8	RPM	1,750	1,750	3,500	1,750	1,750	1,150	3,550	3,550
9	BHP	1-3½	0.1-0.5	40-150	5-20	25-100	7½-30	¶8½-10	10-50
10	$Q_{min/max}$	40-95	2-8	‡220-1,550	‡120-780	900-3,000	1,000-2,500	60-150	100-320
11	NPSHR	4-14	4-18	7-35	6-20	8-25	5-27	7-16	5-25
12	NPSHR/D	No	—	Slightly	Yes	No	No	—	Yes
13	H_{SH}	43-110	‡116-162	‡185-460	50-120	56-145	22-64	¶240	150-440

*Regenerative turbine pump. †"D" constant for group (10, 15, 20) with varying no. of blades. ‡Recommended area of operation indicated by heavy line for figs. 10.11 and 10.12. §Computed. ¶Per stage, as noted.

TABLE 10.27
EVALUATION OF PUMP PERFORMANCE CHARACTERISTICS

	ITEMS	FIG. 10.17	FIG. 10.18	FIG. 10.19	FIG. 10.20	FIG. 10.21	FIG. 10.22	FIG. 10.23	FIG. 10.24
1	D	6-10	19½-21¼	§10-13⅛	§8-13⅛	28-38	22-28	25⁵⁄₁₆-28⅝	A, B, C ¶
2	Q_{max}	110-180	13,000-14,600	900-1,300	900-1,500	30,000-50,000	32,500-42,000	75,000-85,000	**57,500-85,000
3	H	15-105	*‡60-210	63-175	110-710	275-700	375-850	*‡§25-116	**6-30
4	Q_E	80-150	*7,000-14,000	600-1,200	800-1,300	20,000-40,000	20,000-40,000	*40,000-80,000	40,000-80,000
5	E	40-52	*75-85	40-82	30-76	60-89	60-89	*75-87	**68-88
6	Q_{E-M}	125	11,000	1,100	1,200	34,000	32,500	65,000	60,000
7	H_{M-EM}	85	‡135	144	580	630	730	‡65	20
8	RPM	1,750	1180	1,750	3,550	1,185	1,780	500	500
9	BHP	1.5-6	‡280-440	10-50	30-250	1,000-7,000	1,000-7,600	*‡§680-1,372	§150-1,875
10	$Q_{min/max}$	60-160	*7,000-14,000	500-1,100	400-1,300	15,000-38,000	15,000-37,000	*40,000-80,000	40,000-80,000
11	NPSHR	2-6	*†40-60	5-25	6-30	20-70	40-110	†35-70	25-40 ††
12	NPSHR/D	Yes	No	No	No	No	No	No	No
13	H_{SH}	32-108	‡165-215	90-175	230-720	350-720	530-870	‡§77-116	**§67.6-79.8

*Consult manufacturer for min. operating capacity. †Note requirement for submergence ‡Heads indicated for 2-stage pump. §Indicated. ¶Inquire with manufacturer. **Recommended range of operation as indicated. ††Obtain requirement for minimum submergence.

11.

Variation of Pump Curves

A centrifugal pump characteristic curve for a given impeller diameter and speed (RPM) will vary for several reasons:

11.1. IMPELLER DIAMETER CHANGE IN THE SAME CASING

The following relationships apply approximately at equal efficiencies:

$Q_2/Q_1 = D_2/D_1$ 11–01

$H_2/H_1 = (D_2/D_1)^2$ 11–02

$BHP_2/BHP_1 = (D_2/D_1)^3$ 11–03

Note: These relationships are adequate for estimates. In reality, there will be a deviation from the computed values with these formulae. The larger D_2/D_1, the larger the deviation. The closer to maximum efficiency for the diameter, the lesser the deviation.

■ EXAMPLE 11:1

Replace a 5″ diameter impeller with a 7″ one. The initial capacity was 335 GPM, Head = 62 FT, and BHP = 7.8.

$Q_2 = 335 \times 7/5 = 469$ GPM

$H_2 = 62 \times (7/5)^2 = 121.5$ FT of Liquid

$BHP_2 = 7.8\ (7/5)^3 = 21.4$ BHP

Efficiency in both cases is 67.2%.

11.2. SPEED CHANGE (N)

The following relationships apply approximately only for efficiencies of the same value:

$Q_2/Q_1 = N_2/N_1$ 11–04

$H_2/H_1 = (N_2/N_1)^2$ 11–05

$BHP_2/BHP_1 = (N_2/N_1)^3$ 11–06
$NPSHR_2/NPSHR_1 = (N_2/N_1)^m$ 11–07
where m = 1.3 to 2.

■ EXAMPLE 11:2

Increase the speed of a pump from 1,750 RPM to 3,500 RPM. Compute the new values of these parameters:

$Q_1 = 90$ GPM; $H_1 = 55$ FT of Liquid; BHP = 2.15, and NPSHR = 3 FT.

$Q_2 = 90 \times 3500/1750 = 180$ GPM

$H_2 = 55 \times (3500/1750)^2 = 220$ FT of Liquid

$BHP_2 = 2.15 \times (3500/1750)^3 = 17.2$ BHP

$NPSHR_2 = 3 \times (3500/1750)^{1.5} = 8.48$ NPSHR (FT of 60°F water).

The performance curves reflecting the new conditions are shown on fig. 11.01 and fig. 11.02 (not to scale).

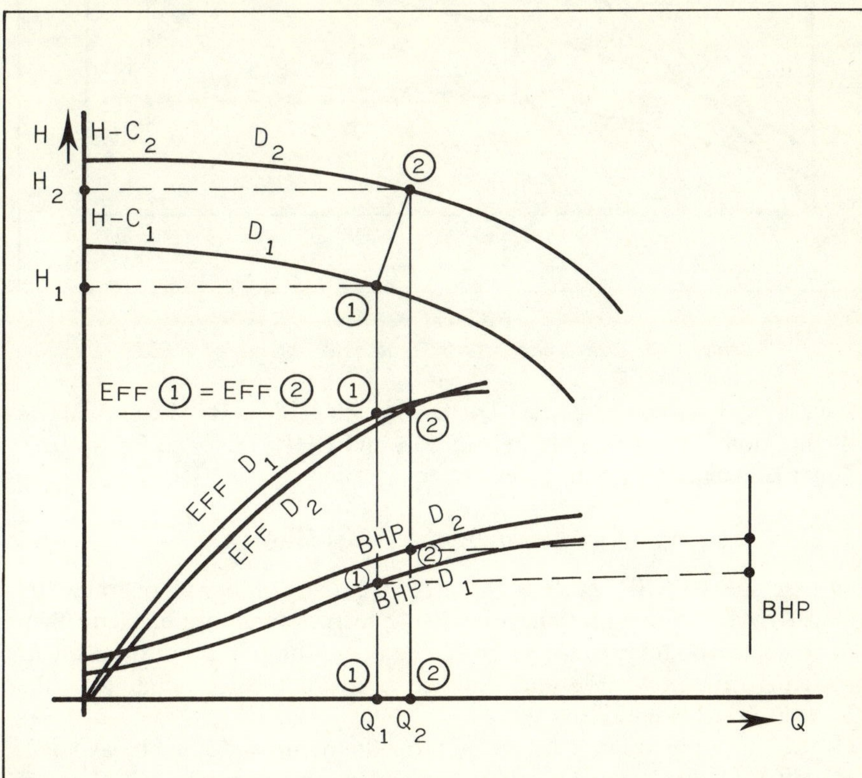

FIG. 11.01

Variation of pump characteristics due to change of impeller diameter.

Condition 1 is represented by D_1, Q_1 H_1; eff D_1, and BHP_1-D_1, and condition 2 by D_2 Q_2, H_2; eff D_2 and BHP_2-D_2. (refer to fig. 11.01)

Note: Efficiencies are equal for both cases.

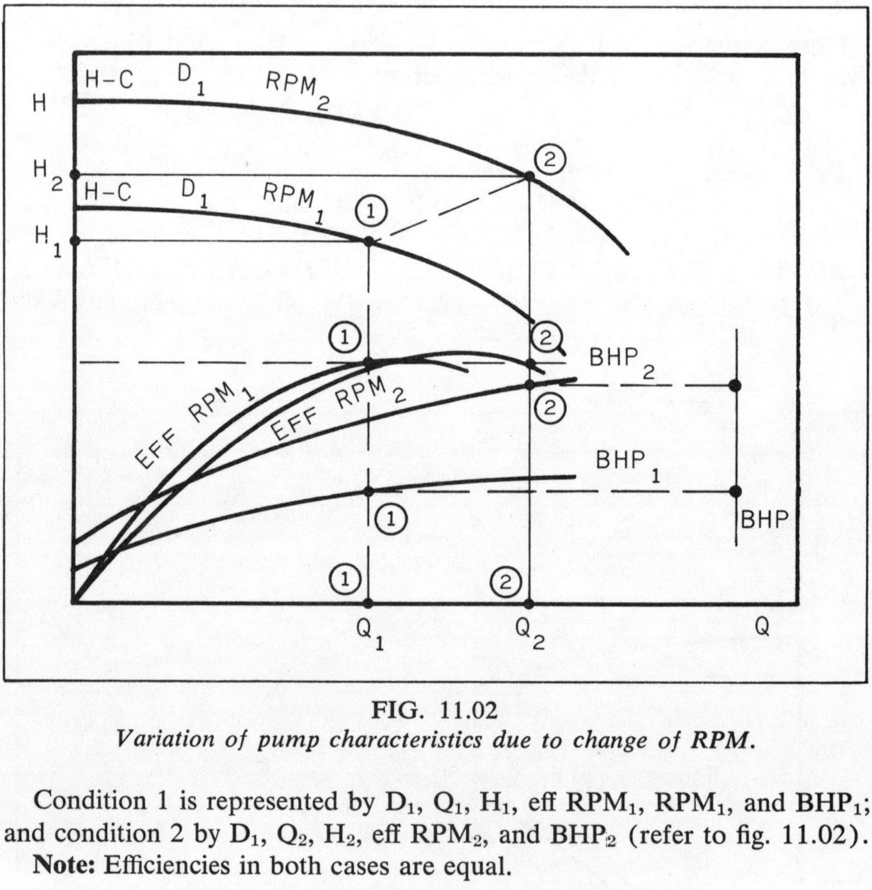

FIG. 11.02

Variation of pump characteristics due to change of RPM.

Condition 1 is represented by D_1, Q_1, H_1, eff RPM_1, RPM_1, and BHP_1; and condition 2 by D_1, Q_2, H_2, eff RPM_2, and BHP_2 (refer to fig. 11.02).

Note: Efficiencies in both cases are equal.

11.3. NPSHA LOWER THAN NPSHR

In case the NPSHR exceeds the NPSHA, the pump will operate "IN THE BREAK." Since the pump is unable to transfer more capacity than the one corresponding to the NPSHA, the operating point will drop off its normal performance curve and stabilize at a point corresponding to the total head for this condition.

Operation in the break is detrimental for the pump and must be avoided. It is said that the pump is "cavitating" when the operating point is off its performance curve.

In figure 11.03, NPSHR and NPSHA are plotted. As shown, for capacity Q_1, NPSHA = NPSHR. It follows that capacities larger than Q_1 wouldn't be transferred by the pump.

H_1 is the corresponding head for this point. As long as H_1 represents the total head for this operation, the operation will remain stable, since the point stays on the pump curve.

In case, for a reason, H_1 drops to H_2, the new operating point will drop from 1 to 2, which is an operation in the break, as defined above. (See chapter 23, page 211).

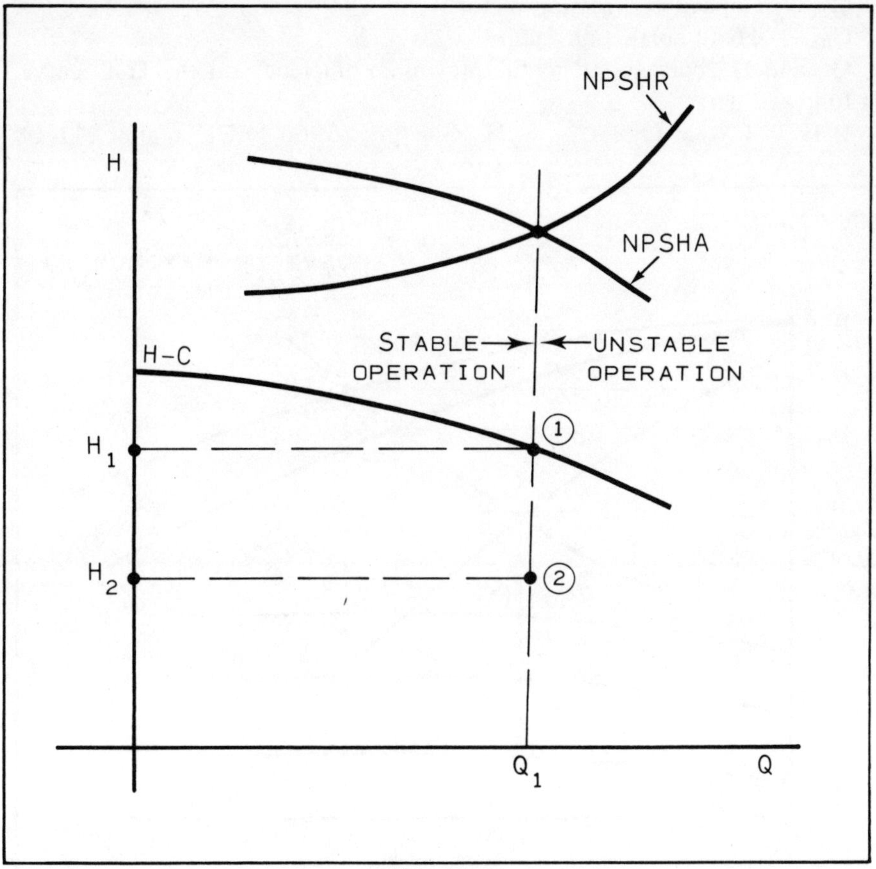

FIG. 11.03
Effect of NPSHR exceeding NPSHA on the operating point of the pump characteristics curve.

11.4 CHANGE OF VISCOSITY

Viscous liquids (200 SSU and above) will affect the performance curve of a centrifugal pump. Both the head and capacity decrease with increasing viscosity; pump efficiency decreases, and BHP increases.

The Hydraulic Institute has published viscosity-correction curves for centrifugal pump performance. These curves permit determination of the correction factors to correct the affected parameters per above from the ones indicated on the manufacturers pump curves which, as said before, refer to water (sp. gr. = 1.0). The factors are read off as a function of pumped capacity, total head, and viscosity.

C_E = Efficiency correction factor

C_{Qm} = Capacity correction factor

C_{Hm} = Head correction factor

Q_m and H_m correspond to the maximum efficiency on the H–C curve. It follows then:

$$BHP = C_{Qm} \times Q \times C_{Hm} \times H \times \text{sp. gr.}/(3960 \times C_E \times \text{eff}) \quad ..11\text{–}08$$

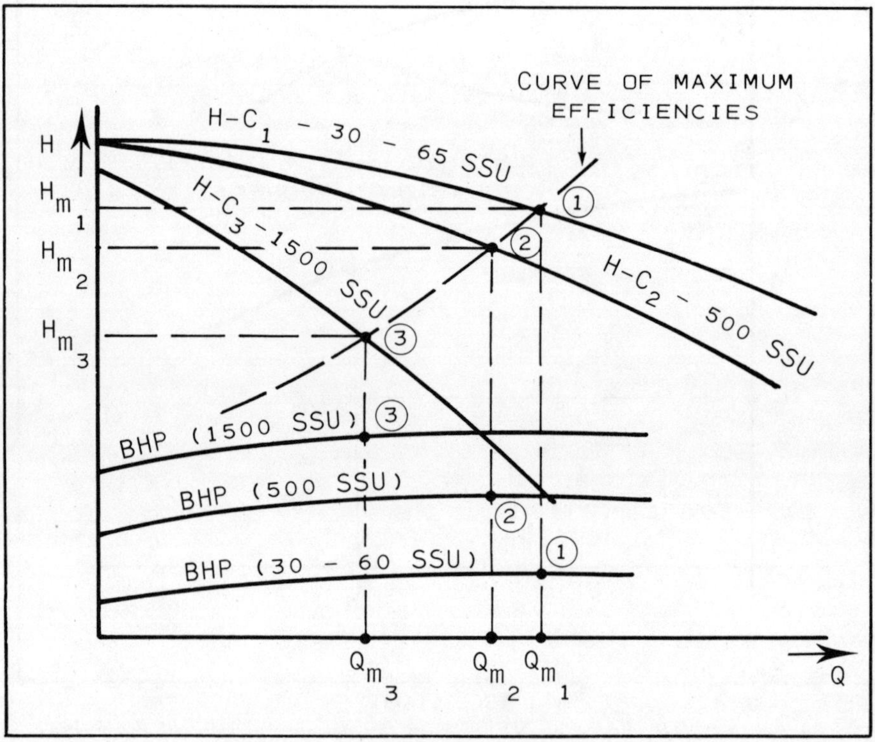

FIG. 11.04
H—C curve variation as function of viscosity.

12.

Specific Speed (NS)

The specific speed of a centrifugal pump is the speed of a geometrically similar pump (a precise geometrical model) at which 1 GPM is delivered at the head of 1 FT of liquid. In practice, specific speed NS is used to link the three main parameters (GPM, H, and RPM) characteristic to the performance of a pump.

The formula for specific speed:

$$NS = RPM \times Q^{0.5}/H^{0.75} \qquad \text{12–01}$$

where Q is in GPM, and H is the head per stage in feet of liquid.

It follows that NS can assume all values between zero and infinity. The only significant value, though, which helps compare different pumps, is the one for the highest efficiency.

Specific speeds range from approximately 500 to 15,000. Centrifugal pumps fall into three major categories: "Volute-type", "Mixed-flow type", and "Axial, or Propeller type". The impeller profiles of these pumps range from radial to axial, and are related to the specific speeds, as fig. 12.01 indicates. Although there is no precise specific speed to separate the three pump categories, one can say that the specific speeds for volute type pumps can range from 500 to 5,000, for mixed-flow type pumps from 4,000–10,000, and for propeller type pumps from 9,000–15,000.

The category, "Volute-type," includes all types of pumps within the range of specific speeds indicated (horizontal and vertical volute-type pumps, vertical turbine pumps, etc.). The majority of the pumps in use fall into this category.

The peak efficiency of a pump and its specific speed are related, as Table 12.02 indicates.

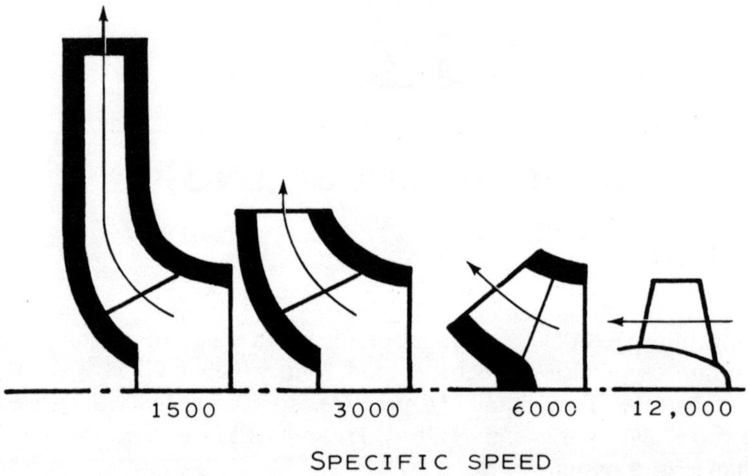

FIG. 12.01
Pump Impeller Profile versus Specific Speed.

TABLE 12.02. APPROXIMATE PUMP PEAK EFFICIENCY RANGE VERSUS PUMP CATEGORY AND SPECIFIC SPEED

PUMP CATEGORY	SPECIFIC SPEED RPM	APPROXIMATE PEAK EFFICIENCY RANGE
Volute	500	45-70%
"	750	55-75%
"	1000	65-85%
"	2000	70-90%
"	3500	82-92%
"	4500	85-90%
Mixed-Flow	4000	85-92%
"	5000	80-90%
"	7000	80-88%
"	10000	80-85%
Axial (Propeller)	9000	82-88%
"	12000	80-85%
"	15000	78-82%

Note that 5,000 RPM and 10,000 RPM are overlapping areas.

TABLE 12.02
Approximate pump peak efficiency range versus pump category and specific speed.

By definition, two pumps are geometrically similar if their specific speeds at optimum efficiency are identical. In this case, the following relationship will be satisfied:

$Q_1/(RPM_1 \times D_1^3) = Q_2/(RPM_2 \times D_2^3)$ 12–02

where Q is in GPM; D in inches.

Low Specific Speed Pumps have flat H–C and BHP curves. The BHP is lower at $Q = 0$ than at rated capacity. The efficiency curve is broad in its best range. Their application: small to average capacities at high heads.

High specific speed pumps have steeper H–C curves, especially in the area of smaller capacities. Beyond a certain point, the operation is not considered stable (lower capacities), and the manufacturers do not advise operation in these areas (Mixed-flow and propeller pumps). The BHP is higher at $Q = 0$ than at rated capacity. The peak efficiency is over a narrow range. The pump driver doesn't necessarily have to be sized for shut-off. In this case, the pump should be started with a partially open discharge valve in order to avoid shut-off pressures.

Application: high capacities at low heads.

■ EXAMPLE 12:1. Compute the specific speed for a pump delivering 4,000 GPM at a head of 100 FT of liquid at a speed of 1,200 RPM. With Equation 12–01: NS = 2,400 RPM. Ans.

■ EXAMPLE 12:2. Compute the specific speed for a pump delivering 20,000 GPM at a head of 60 FT of liquid and a speed of 700 RPM. With Equation 12–01: NS = 4,590 RPM. Ans.

■ EXAMPLE 12:3. The impeller of the pump per example 12:2 has a 16″ diameter. What size should the diameter of the impeller of a geometrically similar pump be for it to deliver 7,000 GPM at the same head of 60 FT of liquid? What is its speed?

Solution: with formulae 12–01, 12–02, and NS = 4,590 RPM from example 12:2:

$$NS = 4{,}590 = \frac{RPM_2 \times Q_2^{0.5}}{H_2^{0.75}} = \frac{RPM_2 \sqrt{7{,}000}}{60^{0.75}} = 3.88\ RPM_2$$

$$RPM_2 = \frac{4{,}590}{3.88} = 1{,}182 \text{ Ans.}$$

$$\frac{20{,}000\ GPM}{700\ RPM \times 16^3} = \frac{7{,}000\ GPM}{1{,}182\ RPM \times D_2^3};\ D_2^3 = 849$$

$D_2 = 9.47''$. Ans.

▪ EXAMPLE 12:4. Determine the speed and impeller diameter for the pump per example 12:3, if it has to deliver 7,000 GPM at a head of 120 FT instead of at 60 FT:

$NS = 4{,}590\ RPM = RPM_2 \times 7000^{0.5}/120^{0.75} = 2.308\ RPM_2;$

$RPM_2 = \dfrac{4590}{2.308} = 1{,}988$ Ans.

With Equation 12–02 and $Q_1 = 20{,}000$ GPM; $RPM_1 = 700$; $RPM_2 = 1{,}988$, $D_1 = 16''$; $D_1^3 = 4{,}096\ in^3$; $Q_2 = 7{,}000$ GPM:

$$D_2^3 = \frac{Q_2 \times RPM_1 \times D_1^3}{Q_1 \times RPM_2} = \frac{7{,}000 \times 700 \times 4{,}096}{20{,}000 \times 1{,}988} = 504.78$$

$D_2 = 7.96$ Ans.

Note: The impeller diameter is smaller (7.96″ vs. 9.47″), but the speed has increased from 1,182 to 1,988 RPM.

13.

Operating Points on H-C Pump Curves

A System-Head Curve (refer to chapters 5 and 7) can be superimposed on the Head-Capacity Curve of a centrifugal pump. The point of intersection is called the OPERATING POINT of the specific pump with the specific system.

Unless the characteristics of the system (System-Head Curve) is changed, the capacity at the operating point will be the only one the pump will deliver for the specific head (See fig. 13.01; refer to item 9.1). This is true, provided the NPSHA exceeds the NPSHR at all times. Otherwise, as indicated before, the operating point will drop off the Head-Capacity Curve, and the pump will be operating in the break.

The four different system curves will make the pump operate at four different operating points: Point 1: Q_1 and H_1; Point 2: Q_2 and H_2; Point 3: Q_3 and H_3; Point 4: Q_4 and H_4. Note that for the H–C pump curve indicated, the head H increases with decreasing capacity Q. For each Q value there is only one H value.

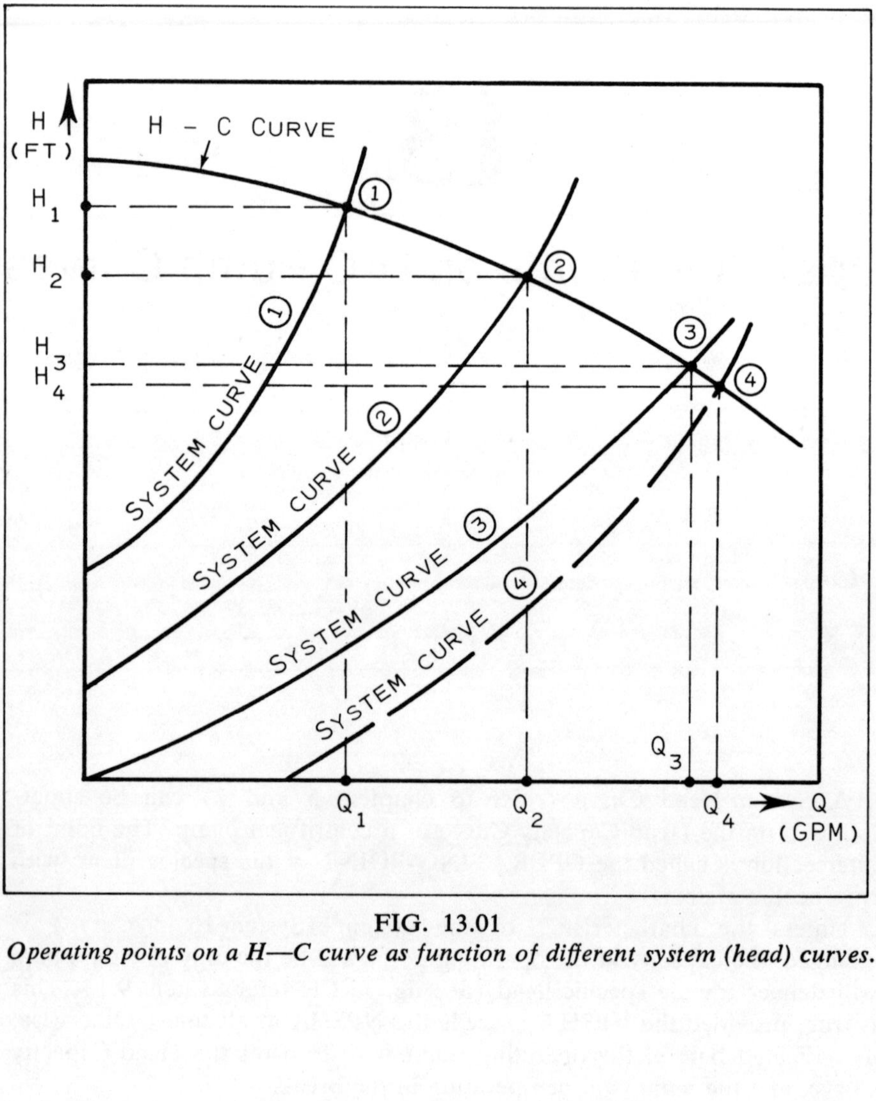

FIG. 13.01
Operating points on a H—C curve as function of different system (head) curves.

14.

Operation in Parallel and in Series

14.1. Two pumps or more, discharging into a common line (common discharge pipe) are said to operate in parallel. For a specific system-head curve, two identical pumps can be selected to discharge double the capacity of each at the same head. Once the pumps are installed, the capacity and the heads will change in case the system-head curve deviates from the anticipated one, or modifications are introduced.

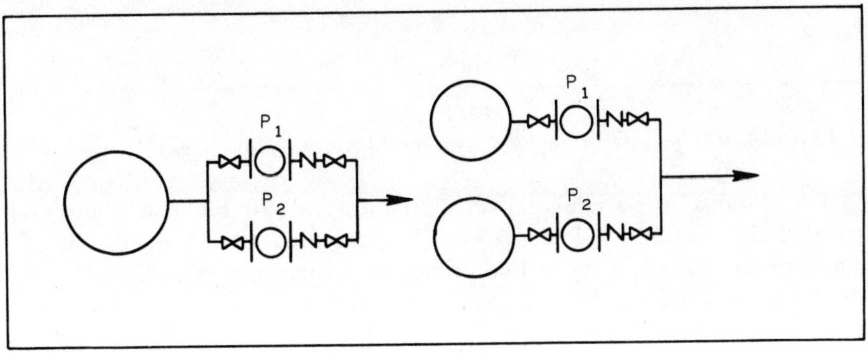

FIG. 14.01
Operation in parallel.

■ EXAMPLE 14:1 (Fig 14.02).

Construct the H-C curve for two identical pumps operating in parallel, superimpose it on different system head curves and comment.

■ OPERATION WITH SYSTEM CURVE "A":

One pump by itself delivers 570 GPM at 54 FT head of liquid.

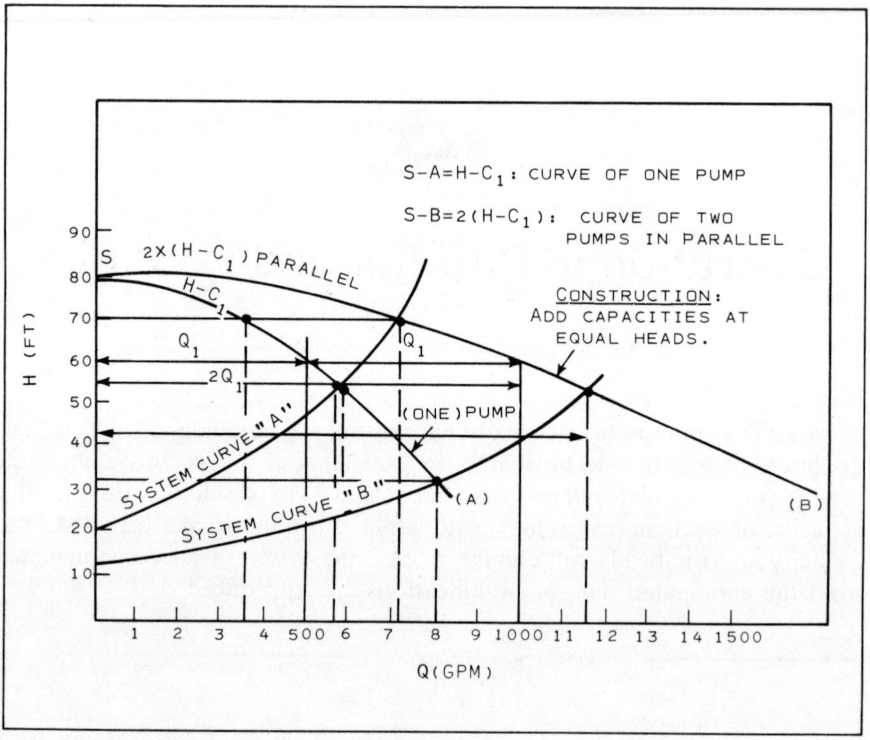

FIG. 14.02

Two identical pumps in parallel. Operating points with curves "A" & "B".

Two pumps in parallel deliver 720 GPM at 70 FT. Each pump in parallel delivers 360 GPM at 70 FT. Compared to the operation of a single pump, the capacity of two pumps in parallel increases by 26%; and the head increases by 29%.

- OPERATION WITH SYSTEM CURVE "B":

One pump by itself delivers 800 GPM at 32 FT of liquid.

Two pumps in parallel deliver 1,150 GPM at 52 FT. Each pump in parallel delivers 575 GPM at 52 FT. Compared to the operation of a single pump, the capacity of two pumps in parallel increases by 44%; and the head increases by 62%.

Conclusion: From the above, it follows that two pumps, operating in parallel, will be more effective with a flatter system head curve. In case two pumps were selected to operate in parallel with system curve "B", the results will be as shown above. A deviation from the computed system curve toward higher heads will reduce the capacity.

In case the system curve is changed from "B" to "A", the total output will drop drastically, from 1,150 GPM to 720 GPM, a reduction of 430 GPM = 430/1150 = 37.4%. The head increase is, in principle, not advantageous. In this case it amounts to 34.6%.

Two pumps in parallel should be selected in case one by itself can not perform as required, or certain flexibility is necessary. Otherwise one pump operating by itself with a spare is, in most cases, the best selection.

14.2 Two pumps or more are said to operate in series when the discharge of the first pump serves as suction for the second pump, and the discharge of the second pump as suction for the third one, etc. The parameters affected are head and capacity.

For a specific system head curve, two identical pumps can be selected to double the head of the first pump. Once the pumps are installed, the capacity and heads will change in case the system head curve deviates from the anticipated one, or modifications are introduced.

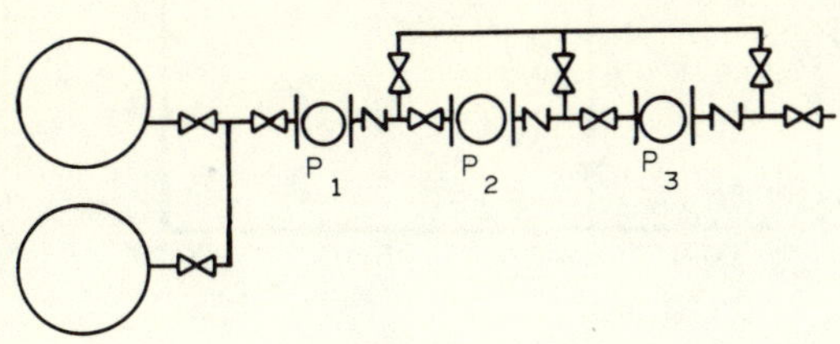

FIG. 14.03
Operation in series.

■ EXAMPLE 14:2 (Fig 14.04)

Construct the H-C pump curve for two identical pumps operating in series, superimpose it on different system head curves and comment.

■ OPERATION WITH SYSTEM CURVE "A":

One pump by itself delivers 1,350 GPM at 225 FT of liquid. Two pumps in series deliver 2,280 GPM at 410 FT. Each pump in series delivers 2,280 GPM at 205 FT. Compared to the operation of a single pump, the capacity of two pumps in series increases by 69%, and the head increases by 82%.

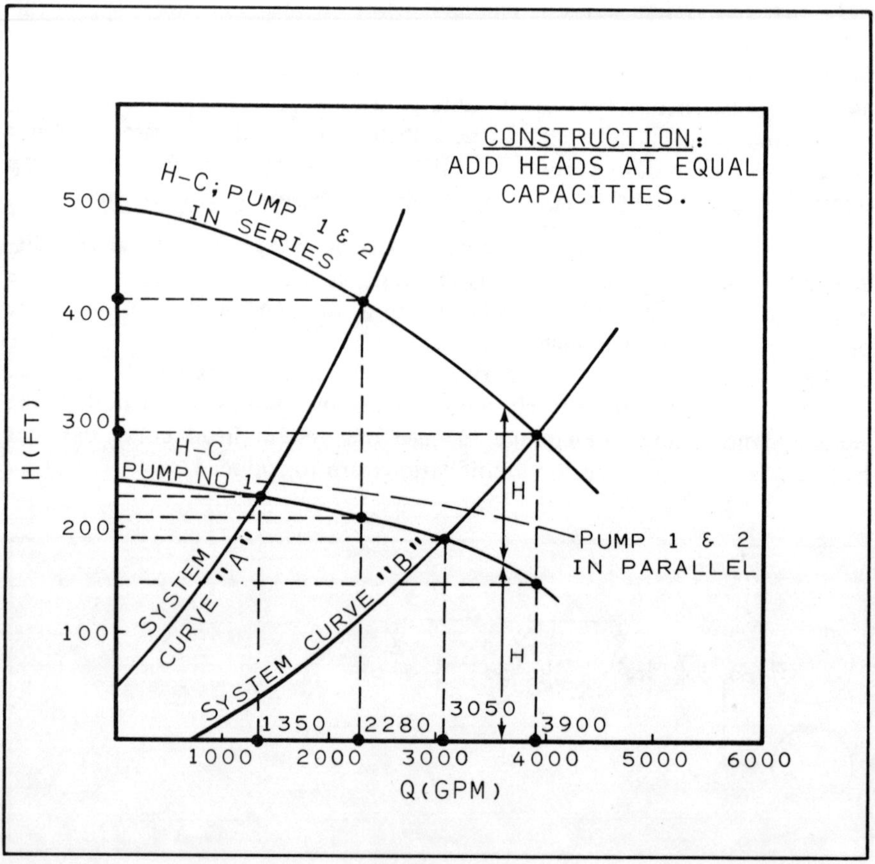

FIG. 14.04
Two identical pumps in series. Operating points with curves "A" & "B".

▪ OPERATION WITH SYSTEM CURVE "B":

One pump by itself delivers 3,050 GPM at 188 FT of liquid. Two pumps in series deliver 3,900 GPM at 286 FT of liquid. Each pump in series delivers 3,900 GPM at 143 FT. Compared to the operation of a single pump, the capacity of two pumps in series increases by 28%, and the head increases by 52%.

The dashed H–C curve on fig. 14.04 represents the H–C curve for the two pumps operating in parallel. As can be seen, the operation with these system curves would be very inefficient. With curve "A" there would, for practical purposes, be no increase in capacity, and only a 10% percent increase with system curve "B".

Conclusion: The operation of two pumps in series with a steeper system curve will be more effective than with a flatter system curve. A deviation from an anticipated system head curve toward higher capacities will reduce the developed head. If the system curve changes from "A" to "B" the head will drop from 410 to 286 FT, a reduction of 124 FT = 124/410 = 30%.

Careful study of the selected pump curves is required to bring about the most efficient and flexible operation.

■ EXAMPLE 14:3

Construct the H-C curve for two pumps with dissimilar characteristics operating in parallel, superimpose it on different system head curves and comment.

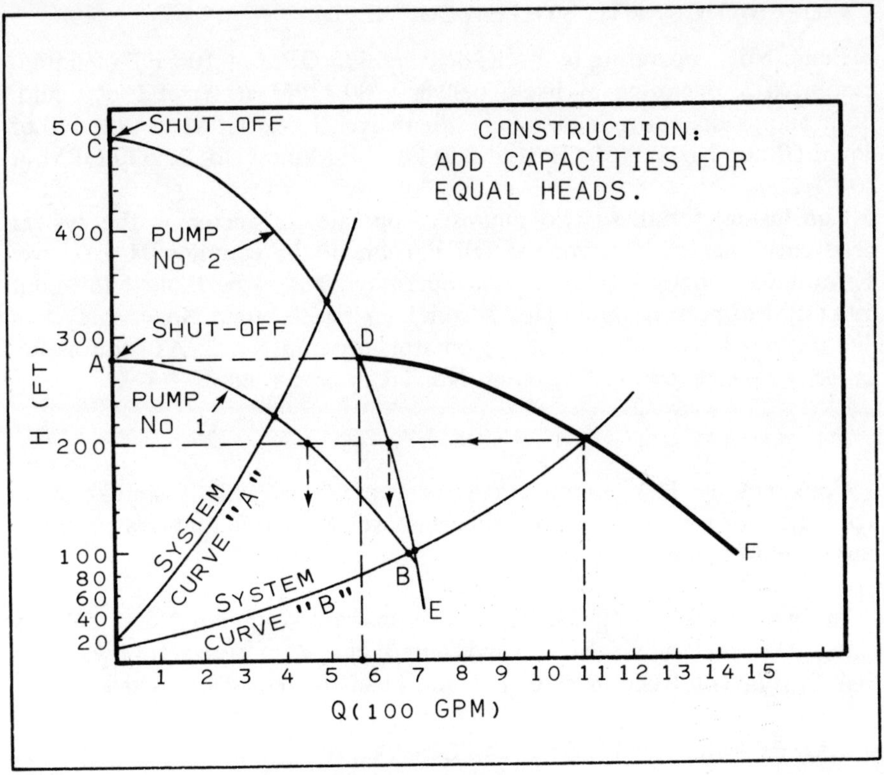

FIG. 14.05
Two pumps with dissimilar characteristics in parallel. Operating points with two different system head curves.

In figure 14.05, pump No. 1 follows curve AB, pump No. 2 follows curve CDE, and pump No. 1 and pump No. 2 in parallel follow curve CDF. For construction of CDF add capacities for equal heads.

■ OPERATION WITH SYSTEM CURVE "A":

Pump No 1, operating by itself, delivers 370 GPM at 225 FT of liquid. Pump No 2, operating by itself, delivers 500 GPM at 332 FT of liquid. With both pumps in parallel, since system curve "A" intersects the common curve CDF before point D, only pump No. 2 will be commanding the transfer of 500 GPM at 332 FT, while pump No. 1 will be backed-off and will operate at shut-off. This represents a very undesirable condition, and pump No. 1 may be damaged. It follows that these two pumps are not suitable to operate in parallel with system curve "A".

■ OPERATION WITH SYSTEM CURVE "B":

Pump No 1, operating by itself, delivers 675 GPM at 100 FT of liquid. Pump No 2, operating by itself, delivers 700 GPM at 103 FT of liquid. Both pumps operating in parallel, will deliver 1,080 GPM at 200 FT of liquid (Pump No. 1: 450 GPM at 200 FT., and Pump No. 2: 630 GPM at 200 FT).

Conclusion: for these two pumps to operate satisfactorily, the system head curve has to intersect the DF portion of the common H–C curve. Throttling, though, during such an operation, has to be limited to about 600 GPM to prevent pump No. 2 from backing-off pump No. 1. A system like this would require to first put on stream pump No. 2. As soon as the capacity exceeds 600 GPM, pump No. 1 can be started.

■ EXAMPLE 14:4

Construct the H–C pump curve for two pumps with dissimilar characteristics, operating in series, superimpose it on different system head curves and comment.

In figure 14.06, pump No. 1 follows curve 1–2, pump No. 2 follows curve 3–5, and pump No. 1 and pump No. 2 in series follow curve 6–4–5. (For construction of 6–4–5 add heads for equal capacities).

■ OPERATION WITH SYSTEM CURVE "A":

Pump No 1, operating by itself, delivers 490 GPM at 185 FT of liquid. Pump No 2, operating by itself, delivers 900 GPM at 340 FT of liquid. Pumps Nos 1 and 2 in series deliver 1,100 GPM at 440 FT of liquid, pump No. 1: 1,100 GPM at a head of 122 FT, and pump No. 2: 1,100 GPM at a head of 318 FT.

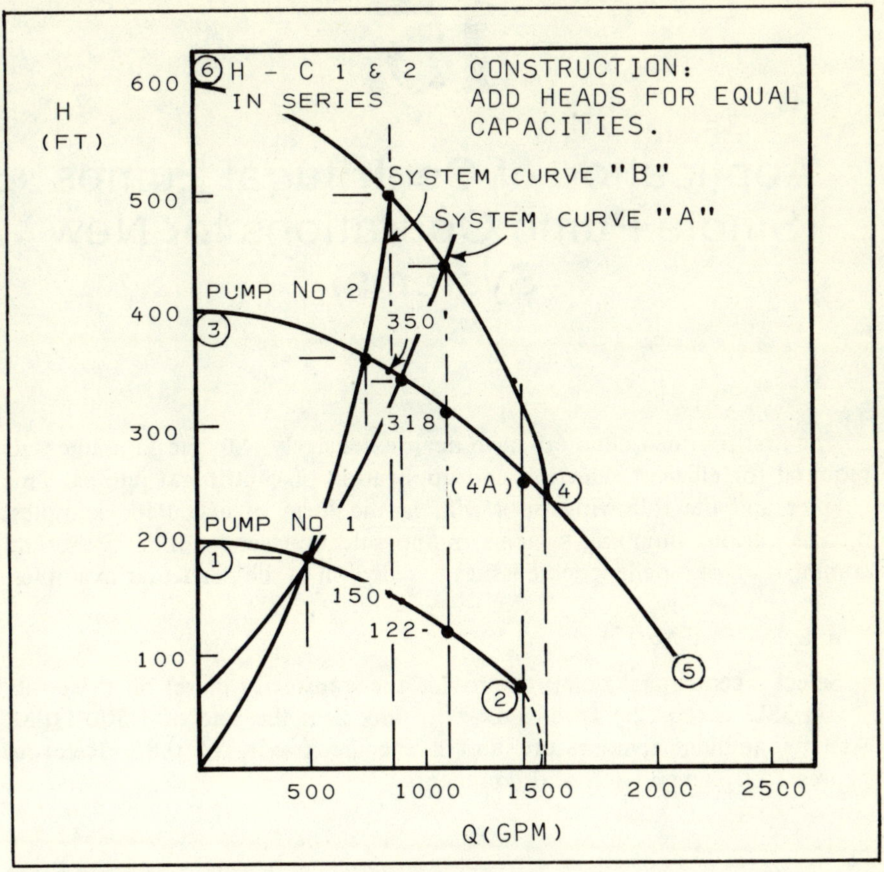

FIG. 14.06

Two pumps with dissimilar characteristics in series. Operating points with two different system head curves.

■ OPERATION WITH SYSTEM CURVE "B":

Pump No 1, operating by itself, delivers 490 GPM at 185 FT of liquid. Pump No 2, operating by itself, delivers 750 GPM at 360 FT of liquid. Pumps Nos 1 and 2 in series deliver 850 GPM at 500 FT of liquid, pump No. 1: 850 GPM at a head of 150 FT, and pump No. 2: 850 GPM at a head of 350 FT.

Note: The operation of pump No. 1 will be unstable beyond point 2. The pump will cavitate. Therefore, operating both pumps in series in the vicinity of point 4 would be detrimental for pump No. 1 and should be avoided. The two pumps would operate very effectively at heads ranging from 300 to 550 FT. In case operation is required between points 4 and 5, only pump No. 2 should be operating.

15.

Application of Centrifugal Pumps (Single Pump Operations for New Systems)

The first fourteen chapters have dealt extensively with the fundamentals required for efficient selection and application of centrifugal pumps. This chapter and the following ones will, in the form of executed examples, discuss various pumping systems, pump selections and applications, and comment on pertinent specific issues touched on in the particular examples.

■ EXAMPLE 15:1

Select a centrifugal pump to provide the transfer of diesel oil (viscosity = 60 SSU = 10 CS) from tank 1 to tank 2 at the rate of 1,500 GPM. Assume ambient temperature and a specific gravity of 0.85. Refer to figures 15.01 and 15.02, and comment.

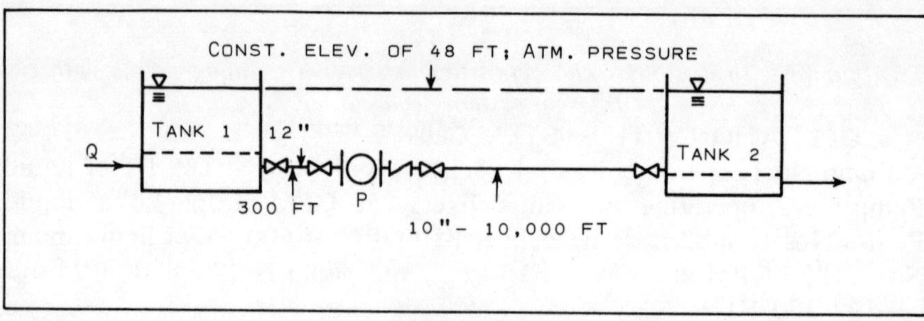

FIG. 15.01

Solution: Since levels are equal and constant, only friction losses will have to be considered. Entrance and exit losses, and losses in equipment like valves and meters, are expressed in equivalent feet and are included in the lengths indicated on fig. 15.01.

Suction:

$RE = 3162 \times GPM/(D'' \times CS) = 3162 \times 1500/(12'' \times 10) = 39{,}525$

$f = 0.184/39{,}525^{0.2} = 0.022$

$$H_{fs} = f \times \frac{Le}{D^5} \times \frac{Q^2}{32.18} = 0.022 \times \frac{300}{12^5} \times \frac{1{,}500^2}{32.18} = 1.86 \text{ FT}$$

Discharge:

$RE = 3162 \times 1500/(10'' \times 10) = 47{,}340$

$f = 0.184/47{,}430^{0.2} = 0.021$

$$H_{fD} = 0.021 \frac{10{,}000}{10^5} \times \frac{1{,}500^2}{32.18} = 147 \text{ FT}$$

$Hf = Hf_S + Hf_D = 1.86 + 147 \approx 149$ FT

Pump operating point is determined at 1,500 GPM @ 150 FT.

Construct the System-Head Curve (Refer to chapter 5).

From above:

$$Hf = Hf_S + Hf_D = 0.022 \frac{300}{12^5} \times \frac{Q^2}{32.18} + 0.021 \frac{10{,}000}{10^5} \times \frac{Q^2}{32.18}$$

$= Q^2/1{,}213{,}250 + Q^2/15{,}323 = Q^2/15{,}132$ FT (Q in GPM).

Points for the System-Head Curve:

Q (GPM)	500	1,000	1,500	2,000	2,500
H (FT)	17	66	149	264	413

Note: The "f" factor is correct for a flowrate of 1,500 GPM only. Deviation resulting from applying the same "f" for all other flowrates is slight and will not affect the shape of the system parabola.

Comments: The selected H–C curve should be, if possible, in the middle of the available H–C curves range. In this case, curve D3 from D1 –D5. The operating point should be in the area of the highest efficiency.

The stable pump operation should extend to a capacity approximately twenty to twenty-five percent higher than the one at the operating point. This is necessary at the start of the operation, when the receiving tank will be empty, and a head of 48 FT will be available. At this time, a transfer rate of about 1,660 GPM can be expected, and the pump should be operating stably. (System-Head curve: $H = Q^2/15{,}123 - 48$).

The H–C curve should not be steep, but should provide the required

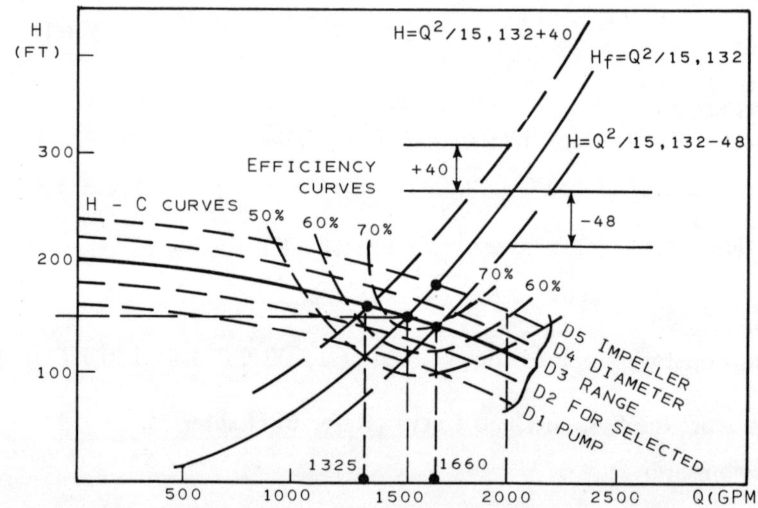

FIG. 15.02
Selected range of H—C curves for given system-head curve.

head in case the level in the suction tank drops to 2–3 feet above the suction nozzle. At this time, the receiving tank may be almost full, and a static head of about 40 feet would be required. The final flowrate would amount to about 1,325 GPM at a head of about 160 FT. (System head curve: $H = Q^2/15{,}132 + 40$)

Replacing the impeller with a larger one in the same casing will provide a larger flowrate for the same system head. For curve D5, 1,650 GPM at 178 FT. (System head curve, $H_f = Q^2/15{,}132$)

The driver should have the power for the maximum capacity the pump can transfer. This would correspond to 2,000 GPM at 155 FT. For an estimated pump efficiency at this point of 60%, we obtain a BHP = $2{,}000 \times 155 \times 0.85/(3{,}960 \times 0.6) = 111$.

The NPSHR should not exceed the NPSHA at any of the considered operation points.

- EXAMPLE 15:2 (Fig 15.03)

Select a centrifugal pump to transfer water at 60°F at the rate of 8,000

GPM from tank 1 to tank 2. The pump operation shall be stable for the maximum level in the suction tank, 48 feet above pump centerline. Establish the operating point for a level in the suction tank 25 feet above the pump centerline.

Compute the flowrates resulting from free gravity flow (by-passing the pump) for liquid levels in suction tank at 48 ft., 25 ft., and 5 ft. above the centerline of the pump.

Compute the maximum distance between the suction tank and the pump which will allow the pump to operate stably for a level of 5 FT in the suction tank and a flowrate of 6,000 GPM. Consider a NPSHR of 30 feet for 60°F water.

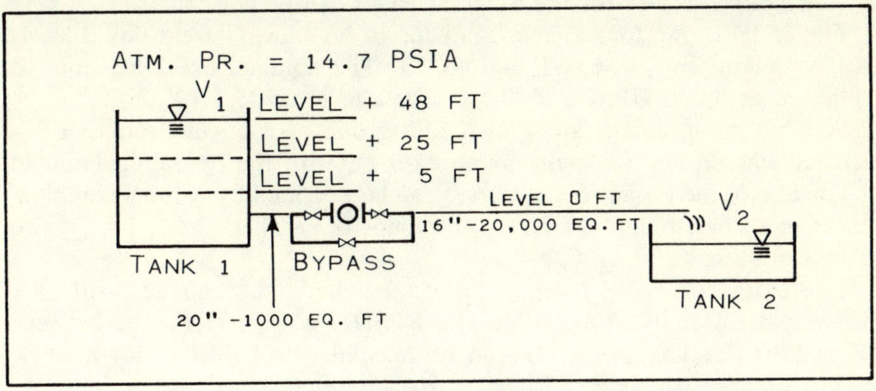

FIG. 15.03

Solution:

Pump Energy:

$$P_1 + Z_1 + V_1^2/2g + Ep - Hf = P_2 + Z_2 + V_2^2/2g$$

where:

$$V_1 = 0;\ P_1 = P_2;\ V_2 = 8{,}000/(2.448 \times 16^2) = 12.76 \text{ FPS},\ Z_2 = 0;$$

$$V_2^2/2g = 2.52 \text{ FT};\ Z_1 = 25 \text{ FT};\ Q = 8{,}000 \text{ GPM}.$$

$$25 + 0 + Ep - H_f = 0 + 2.52$$

$$Ep = Hf - 22.48$$

$$H_f = f_1 \times \frac{L_1}{D_1^5} \times \frac{Q^2}{32.18} + f_2 \times \frac{L_2}{D_2^5} \times \frac{Q^2}{32.18}$$

$$H_f = \frac{Q^2}{32.18}\left[f_1 \frac{L_1}{D_1^5} + f_2 \frac{L_2}{D_2^5} \right]$$

$$RE_1 = \frac{3162 \times 8{,}000}{20 \times 1} = 1{,}264{,}000; \; RE_2 = \frac{3162 \times 8{,}000}{16 \times 1} = 1{,}581{,}000$$

For new pipes, $f_1 = 0.013$, and $f_2 = 0.012$

$$H_f = \frac{8000^2}{32.18}\left[0.013\frac{1{,}000}{20^5} + 0.012\frac{20{,}000}{16^5}\right] = 463.28 \text{ FT of water}$$

$$Ep = H_f - 22.48 = 440 \text{ FT}$$

$H_f = Q^2/138{,}148$ FT—friction loss system curve, from above.

The operating point for a 25 FT liquid level in the suction tank is 8,000 GPM at 440 FT. Ans.

Now system curves for the different levels can be constructed.

Fig. 15.04 shows four system-head curves for four different liquid levels in the suction tank: 0, 5, 25, and 48 FT. The nominal operating point is indicated as 8,000 GPM at 440 FT (for a level of 25 FT).

Suitable pump curves are shown as a group of five (five different impeller diameters for the same pump casing), with the operating point in the middle of the group. These curves can be obtained by various combinations: one single pump ($H–C_1$); two pumps in parallel ($2 \times H–C_2$), two pumps in series ($2 \times H–C_3$).

Due to the dropping level in the suction tank, the capacity will vary between 8,200 and 7,800 GPM. The NPSHR is expected to be between 25 and 30 FT. The motor HP can be calculated to be 1,050 for a single pump, or approximately 550 HP for each of the two pumps in series or in parallel.

Comments: A single pump per $H–C_1$ would be very efficient, and operation would be easier. In case a spare pump is required, the cost will go up.

Two pumps in parallel ($2 \times H–C_2$) could operate satisfactorily. They would be more expensive than one large pump, but less expensive than two large ones. The disadvantage in this case is that one pump cannot operate by itself stably, since its curve doesn't intersect the system head curve. Therefore a third pump would be required in case of emergency. This would make the system very expensive. Therefore two pumps in parallel are not suitable for this case. Also, to operate two pumps in parallel is more involved than operating a single pump.

Two pumps in series ($2 \times H-C_3$) will produce curve $H-C_1$. The advantage in this case is that one pump can operate stably and deliver about 6,000 GPM. If the process permits this decrease, it would represent the least expensive and most flexible system. If not, then one single pump with a spare would be the best choice. The designer has to weigh all alternatives, cost, and arrangement before making a final selection.

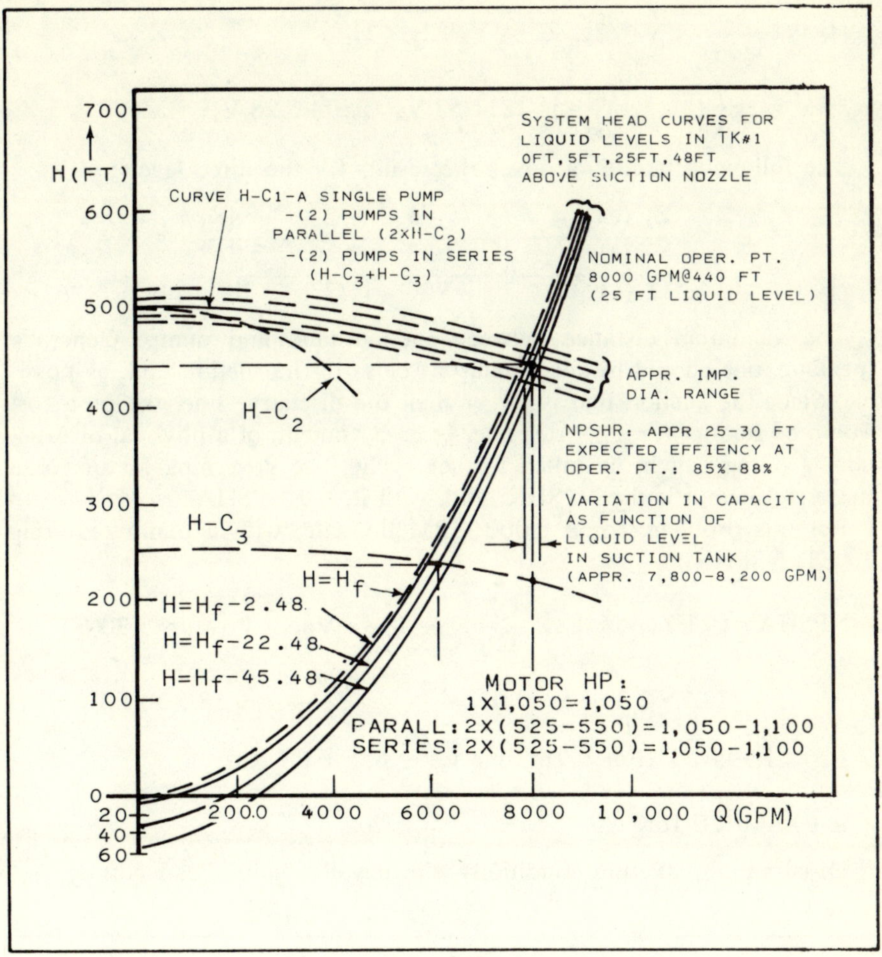

FIG. 15.04
Single pump operation vs two pumps in parallel or series to satisfy required operating point.

Computation of Gravity Flow (pump by-passed), for the existing conditions:

$L_e = 20{,}000$ FT; $f = 0.014$; $D = 16/12 = 1.33$ FT;
(suction line neglected); $P_1 = P_2$; $Z_2 = 0$

$Q = 352.6 \times D^2 \text{ (FT)} \times V_{(FPS)}$, from 1–20

$$P_1 + Z_1 + V_1^2/2g - H_f = P_2 + Z_2 + V_2^2/2g$$

$$Z_1 + O - H_f = O + V_2^2/2g$$

$$H_f = f\frac{Le}{D_{(FT)}} \times V_2^2/2g;\ Z_1 = V_2^2/2g + H_f$$

$$= V_2^2/2g\left(1 + f\frac{Le}{D_{(FT)}}\right) = 211.53\ V_2^2/2g = 3.28\ V_2^2 = Z_1$$

The following tabulation gives the results for the three levels:

Z_1 (FT)	5	25	48	
V_2 (FPS)	1.23	2.76	3.82	
Q (GPM)	767	1,722	2,382	Ans.

The maximum distance between suction tank and pump: Generally speaking, one should install a pump as close to the suction tank as possible. Since the suction line is larger than the discharge line, the pipe cost would be smaller for a shorter line. In case, though, of a physical obstruction, a pump can be installed farther away. The governing factor for a successful layout is the NPSHR; and, with it, the NPSHA.

For this problem, for a 6,000 GPM flowrate, with a pump requiring 30 FT. NSPH:

$$NPSHA = P_1 + Z_1 - f \times Le \times \frac{Q^2}{D^5} \times \frac{1}{32.18} - \text{Vapor Pressure} > 30\ FT$$

$$= 14.7 \times 2.31 + 5 - 0.012\frac{Le}{20^5} \times \frac{6{,}000^2}{32.18} - 0.58 > 30\ FT$$

$Le < 1{,}995$ FT (For a 16″ line $Le < 655$ FT). Ans.

■ EXAMPLE 15:3

Discuss pump suction conditions with negative static head (lift).

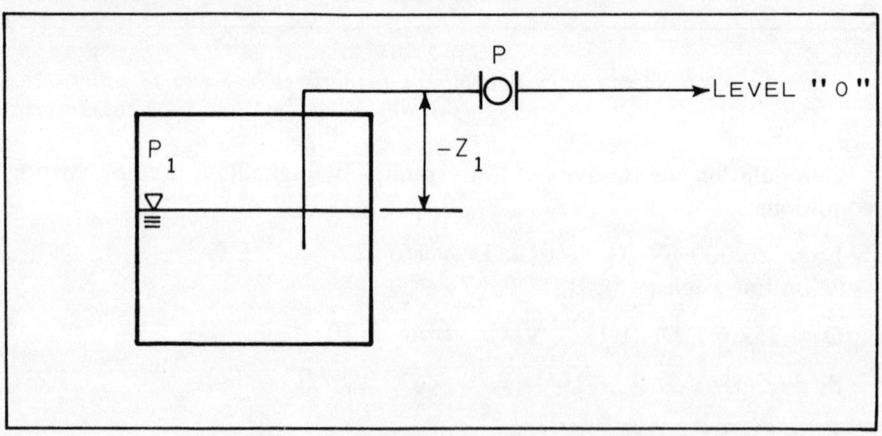

FIG. 15.05

An installation per fig. 15.05 can perform satisfactorily only if the NPSHA is larger than the NPSHR which is the requirement for any pump installation.

$NPSHA = P_1 + (-Z_1) - Hfs - \text{Vapor pressure } (V_P) > NPSHR$

From this equation it follows that, if the liquid is boiling, or close to boiling, P_1 equals to the Vapor pressure (Refer to chapter 1.14). Then $NPSHA = -Z_1 - H_{fs} < 0$, i.e., a horizontal pump cannot transfer boiling liquid from an elevation below the pump centerline; the NPSHA must be positive and larger than NPSHR.

Consider the following conditions for non-boiling conditions: NPSHR = 10 FT; P_1 = 34 FT (open tank, 14.7 PSIA); H_{fs} = 5 FT suction loss; V_P = 0.6 FT (water at 60° F)

$$10 = 34 - Z_1 - 5 - 0.6$$

$$Z_1 \leqq 18.4 \text{ FT.}$$

The level in the tank must not be more than 18.4 FT below the pump centerline.

For NPSHR = 12 FT, P_1 = 14.7 PSIA = 45 FT of gasoline (gasoline, 60°F, sp. g. 0.75), H_{fs} = 5 FT, and V_P = 2.5 PSIA = 7.7 FT:

$$12 = 45 - Z_1 - 5 - 7.7$$

$$Z_1 \leqq 45 - 12 - 5 - 7.7 \leqq 20.3 \text{ FT (below centerline of pump).}$$

For design purposes, use 80% of 20.3 = 16.25 FT.

For NPSHR = 10 FT, P_1 = 14.7 PSIA = 35.1 FT (water at 190°F, sp. g. = 0.967), H_{fs} = 4 FT, and Vp = 9.34 PSIA = 12.8 FT.

$$10 = 35.1 - Z_1 - 4 - 12.8$$

$$Z_1 \leqq 35.1 - 10 - 4 - 12.8 \leqq 8.3 \text{ FT (max. below centerline)}$$

For design purposes: $Z_1 = 0.8 \times 8.3 = 6.5$ FT.

Conclusion: Suction conditions with negative static head (lift) have to be investigated very carefully in order to avoid cavitation. When possible, vertical pumps should be used which are submerged in the liquid and have, thus, a positive suction head.

■ EXAMPLE 15:4

Determine the minimum static head required for the delivery of 4,000 GPM of boiling water at 212°F (14.7 PSIA) for a NPSHR of 20 FT. Refer to fig. 15.06.

$NPSHA = P_1 + Z_1 - Hfs - V_p$, where $P_1 = V_p$ (boiling water);

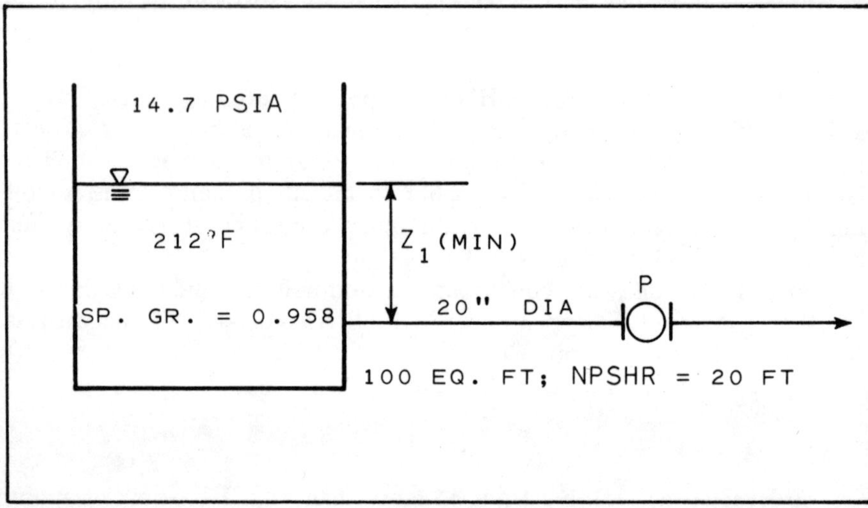

FIG. 15.06

$$= Z_1 - Hfs = Z_1 - f\frac{Le}{D^5}\times\frac{4{,}000^2}{32.18};\ \text{assume } f = 0.02.$$

Then:

$$NPSHA = Z_1 - 0.02\frac{100}{20^5}\times\frac{4{,}000^2}{32.18}$$

$$NPSHR = 20;$$

$$20 = Z_1 - 0.31$$

$$Z = 20.31 \text{ FT}$$

For design purposes: $20.31/0.8 \approx 25$ FT—Ans.

■ EXAMPLE 15:5

Determine the operating point for a condensate pump, delivering water condensate from a condenser to a feed-water heater. For arrangement, refer to fig. 15.07. Suction friction losses = 1 FT; discharge friction losses = 15 FT, Capacity = 400 GPM, atmospheric pressure = 14.7 PSIA, heater pressure = 15 PSIG, pressure in condenser = 2 in. of mercury, and temperature = 101.5° F. The pump NPSHR = 6 FT, $V_1 = V_2 = 0$, $P_1 = \frac{2''}{30''}\times 14.7 = 0.98$ PSIA $= 0.98\times 2.31/0.95 = 2.38$ FT ABS, $Z_1 = 10$ FT; $Z_2 = 50$ FT, $H_f = H_s + H_D = 1 + 15 = 16$ FT, $P_2 = 15 + 14.7 = 29.7$ psia

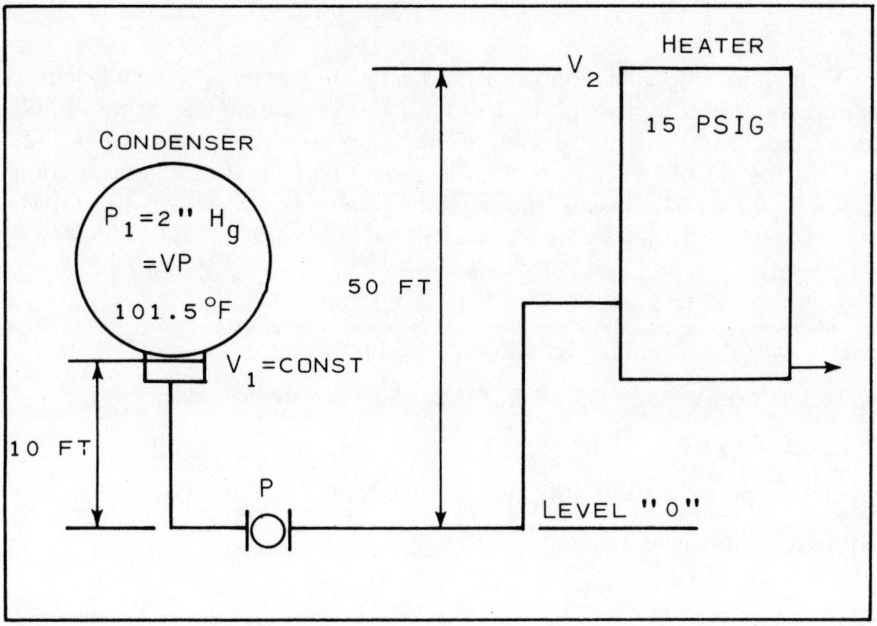

FIG. 15.07

$=29.7\times2.31/0.95=72.22$ FT ABS, and Vapor Pressure$=P_1$.

$$P_1+Z_1+V_1^2/2g-H_f+E_p=P_2+Z_2+V_2^2/2g$$

$$2.38+10+0-16+E_p=72.22+50+0$$

$$E_p=125.84 \text{ FT}$$

Operating point=400 GPM at 126 FT Ans.

$$\text{NPSHA}=P_1+Z_1-H_{fs}-V_p;\ P_1=V_p$$

Thus:

$$\text{NPSHA}=Z_1-H_{fs}=10-1=9 \text{ FT}>6 \text{ FT for NPSHR.}$$

Comments: When pumping hot water, one should know the correct NPSHR for the specific temperature. NPSHR on pump curves relates to water at 60° F. It will increase for higher temperatures. Water at 250° F requires 2–3 FT more than the one indicated. One should consult the manufacturer.

■ EXAMPLE 15:6

A pump will take suction from the heater in example 15.5 and deliver boiling water at 29.7 psia (250° F) to a boiler generating a pressure of 500 psig. Compute the operating point of the pump for a capacity = 400 GPM. Refer to fig. 15.08. Suction friction losses = 10 ft, and discharge friction losses = 40 ft. The loss at the regulating valve is 20 psi for 400 GPM, $H_f = H_{fs} + H_{fD} = 10 + 40 = 50$ FT, loss in valve $Hv = 20 \times 2.31/0.95 = 48.6$ FT, total loss=98.6 FT, $P_1 = 29.7$ psia=72.22 FT, $Z_1 = 50$; $Z_2 = 40$, $P_2 = (500 + 14.7) \times 2.31/0.95 = 1251.53$ FT, and $V_1 = V_2 = 0$.

$$P_1 + Z_1 + V_1^2/2g - H_f - H_v + E_p = P_2 + Z_2 + V_2^2/2g$$

$$72.22 + 50 + 0 - 50 - 48.6 + E_p = 1251.53 + 40 + 0$$

$$E_p = 1267.91 \text{ FT}$$

Operating point is 400 GPM at 1,275 FT. Ans.

NPSHA in this case is ample.

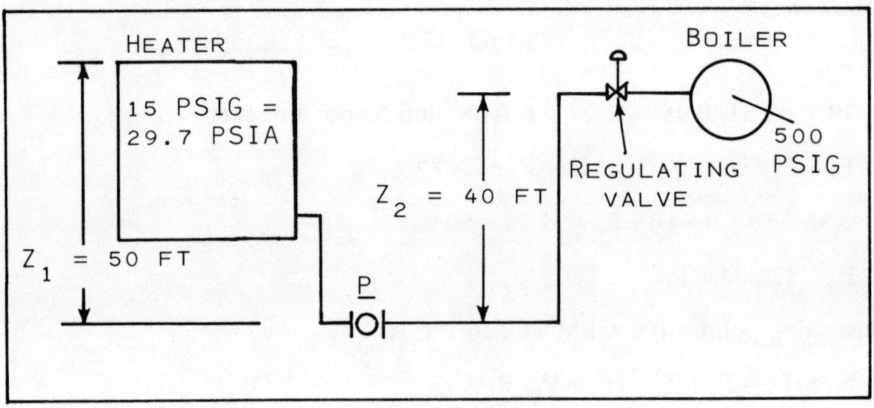

FIG. 15.08

■ EXAMPLE 15:7

Select a pump to transfer product from one vessel to another. The level in the receiving vessel varies. The pump should operate stably and efficiently at all indicated levels for it. Capacity = 800 GPM, sp.gr. = 1.2; NPSHA is ample; viscosity=2CS=32 SSU; $P_1 = 20$ PSIG (constant); $P_2 =$ 60 PSIG (constant); and atm. pressure=14.7 PSIA. Refer to fig. 15.09.

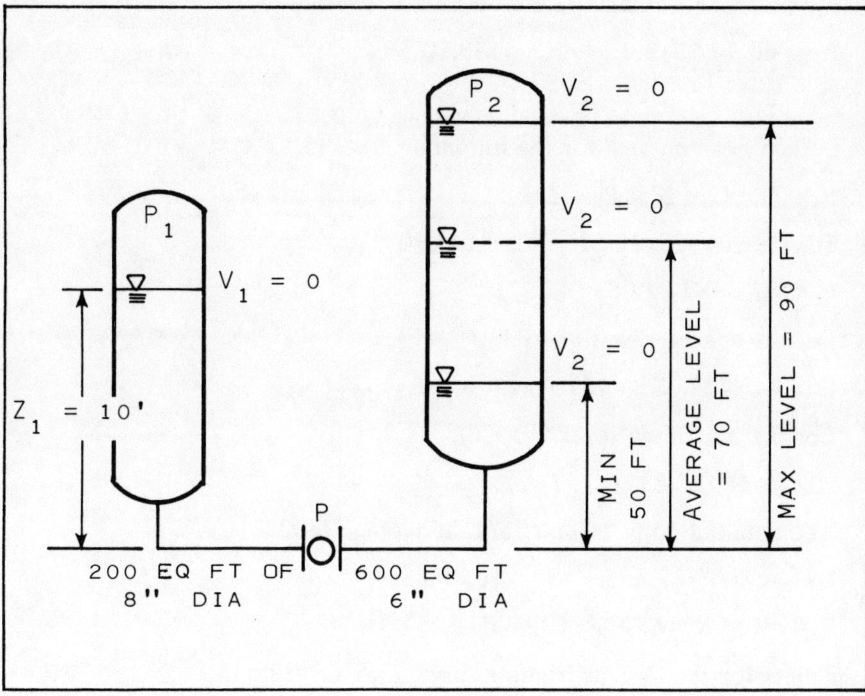

FIG. 15.09

Compute and plot the system head curve. For the suction:

$$RE_s = \frac{3162 \times Q\ (GPM)}{D(in.) \times CS} = \frac{3162 \times 800}{8 \times 2} = 158{,}100$$

fs=0.018 (from chart for new steel pipes)

$$Hs = fs\frac{Ls}{D_s^5} \times \frac{Q^2}{32.18} = \frac{0.018 \times 200}{8^5 \times 32.18} \times Q^2 = Q^2/293{,}000$$

For the discharge:

$$RE_D = \frac{3{,}162 \times 800}{6 \times 2} = 210{,}800$$

fd=0.017 (from charts)

$$H_D = \frac{0.017 \times 600}{6^5 \times 32.18} \times Q^2 = Q^2/24{,}530$$

$$H_f = H_s + H_D = Q^2\left(\frac{1}{293{,}000} + \frac{1}{24{,}530}\right) = Q^2/22{,}635 = \text{System friction}$$

curve. Ans.

$P_1=(20+14.7)\times 2.31/1.2=66.80$ FT

$P_2=(60+14.7)\times 2.31/1.2=143.80$ FT

$V_1=V_2=0$

System head curves: for the minimum level (50 FT)=

$P_1+Z_1-H_f+E_p=P_2+Z_2$

$66.80+10-Q^2/22{,}635+E_p=143.80+50$

$E_p=H_{Min}=Q^2/22{,}635+117$

For the average level (70 FT):

$H_{AV}=Q^2/22{,}635+137$ Ans.

For the maximum level (90 FT):

$H_{MAX}=Q^2/22{,}635+157$ Ans

The nominal point of operation at average level:

($H_{AV}=800^2/22{,}635+137=165$ FT)

800 GPM at 165 FT of liquid. Ans.

Points for the system friction curve, $H_f=Q^2/22{,}635$:

Q (GPM)	200	400	600	800	1,000
H_f (FT)	1.8	7	16	28	44

Note: "f" values, good for 800 GPM, were used for the whole range. The slight difference may be neglected.

Refering to fig. 15.10, consider these alternatives:

Alternative 1: Pump curve H–C_1 satisfies the operating point (800 GPM at 165 FT). The operating point for H_{min} = 920 GPM at 153 FT (point 4), and operating point for H_{max} = 665 GPM at 178 FT (point 1).

Variation from nominal operation point for capacity is from 83% – 115%; and for head is from 92% – 108%.

Alternative 2: Pump curve H–C_2 satisfies the nominal operating point (800 GPM at 165 FT). The operating point for H_{min} = 855 GPM at 149 FT (point 2), and operating point for H_{max} = 730 GPM at 182 FT (point 3).

Variation from nominal operating point for capacity is from 91% – 107%, and for head is from 90% – 110%.

Note: To obtain the operating point for intermediate levels in vessel 2, say at 80 ft, draw a parallel H curve ($H_{80\ ft}=H_f+147$) and read off fig. 15.10, 735 GPM on curve H$-C_1$, and 765 GPM on curve H$-C_2$.

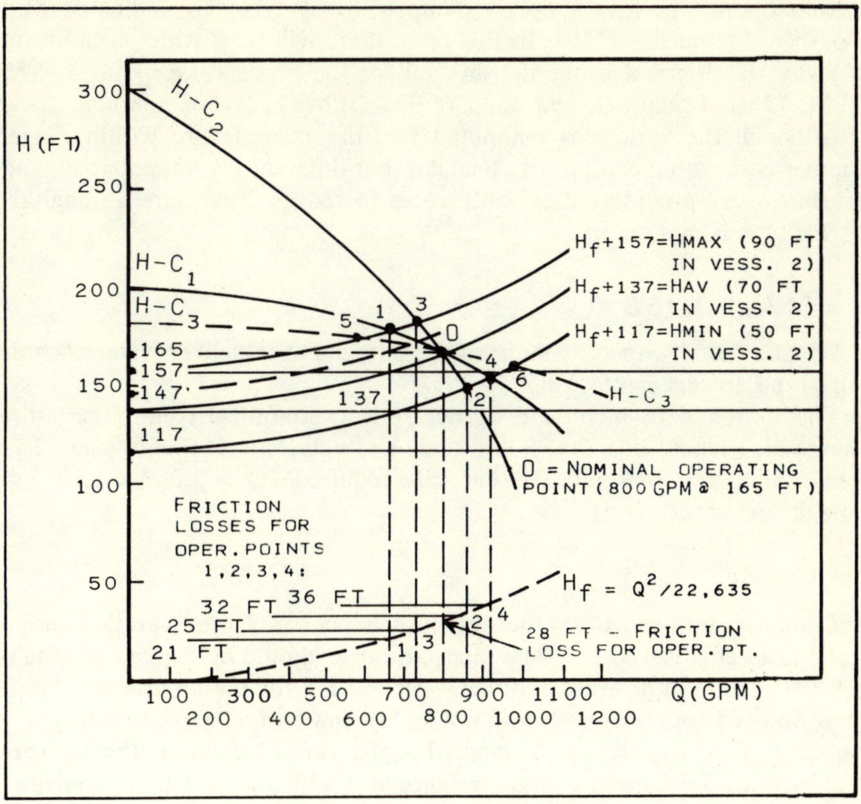

FIG. 15.10
Alternative solutions for example 15:7.

Comments: The advantages of curve $H-C_1$ are low shut-off pressure and low pressure variation between points 1 and 4. The disadvantage of curve $H-C_1$ is a large capacity variation between points 1 and 4.

The advantages of curve $H-C_2$ are low capacity variation between points 3 and 2; and the ability to attain higher heads in case the system head curve changes in the future (higher static heads).

The disadvantages of curve $H-C_2$ are a high shut-off head and a large head variation between points 3 and 2.

Conclusion: The application engineer will have to make the final selection based on specific process requirements, future expansion, pump efficiency, type of operation, etc. To just specify the nominal operating point would result in inefficient pump operation.

The manufacturer must know about the level variations in vessel 2. In case he doesn't, he may supply a pump with curve $H-C_3$, indicated with a dashed line on fig. 15.10. In this case, there will be a wide variation in capacity which could prove detrimental for the process, i.e. point 5, 575 GPM, 72% of nominal; and point 6, 975 GPM, 122% of nominal.

Although the variations computed for this example are within range, another case with a similar arrangement, but different layout, capacity, and pressures may provide critical differences in ranges. Therefore an analysis of this type is mandatory.

■ EXAMPLE 15:8

Select a pump with suitable head-capacity curve to deliver water according to the arrangement on fig. 15.11.

The design data include a vertical, electric-motor-driven pump; discharge pipe length of 6,000 FT; fittings and valves = 1,500 equivalent FT; total length (L_{eq}) = 7,500 FT; capacity required (Q) = 3,000 GPM; and pipe diameter (D) = 10″.

Solution:

Compute and construct the system-head curve. Note that the design static head equals 100 FT. The pump, though, should be capable to establish the syphon at a level of 400 FT, or 600 FT for the alternative. Minimum velocity for start-up should not be less than 6 fps, corresponding to a capacity of Q_s = 1,500 GPM. Note that the control valve at the delivery tank will regulate the discharge pressure at it which can be high, considering the static elevation (H).

For the arrangement in figure 15.11, $P_1 = P_2$, $Z_1 = 0$, $Z_2 = 100$ FT, $V_1 = V_2 = 0$, RE $\approx$ 1,000,000, and f = 0.014 (new steel pipes).

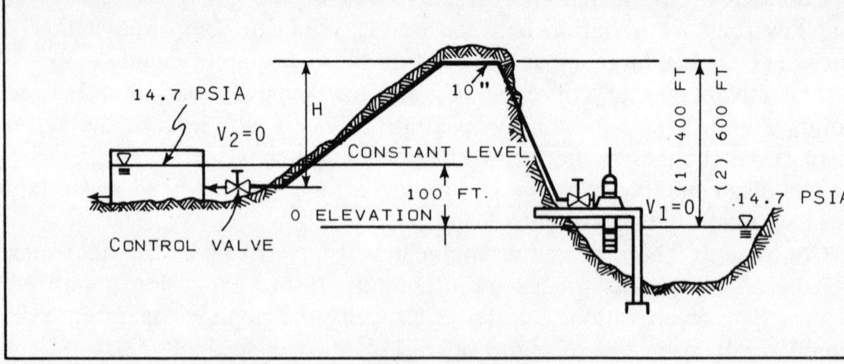

FIG. 15.11

$$P_1 + Z_1 + V_1^2/2g - H_f + E_p = P_2 + Z_2 + V_2^2/2g$$

$$E_p = 100 + H_f = 100 + f \times \frac{Le}{D^5} \times \frac{Q^2}{32.18} = 100 + 0.014 \frac{7500}{10^5} \times \frac{Q^2}{32.18}$$
$$= 100 + Q^2/30{,}650 \text{ Ans.}$$

Values for H_f and E_p:

Q (GPM)	500	1,000	1,500	2,000	2,500	3,000
H_f	8	33	73	130	204	294
$H=E_P$	108	133	173	230	304	394

f assumed constant for the above range. Refer to fig. 15.12.

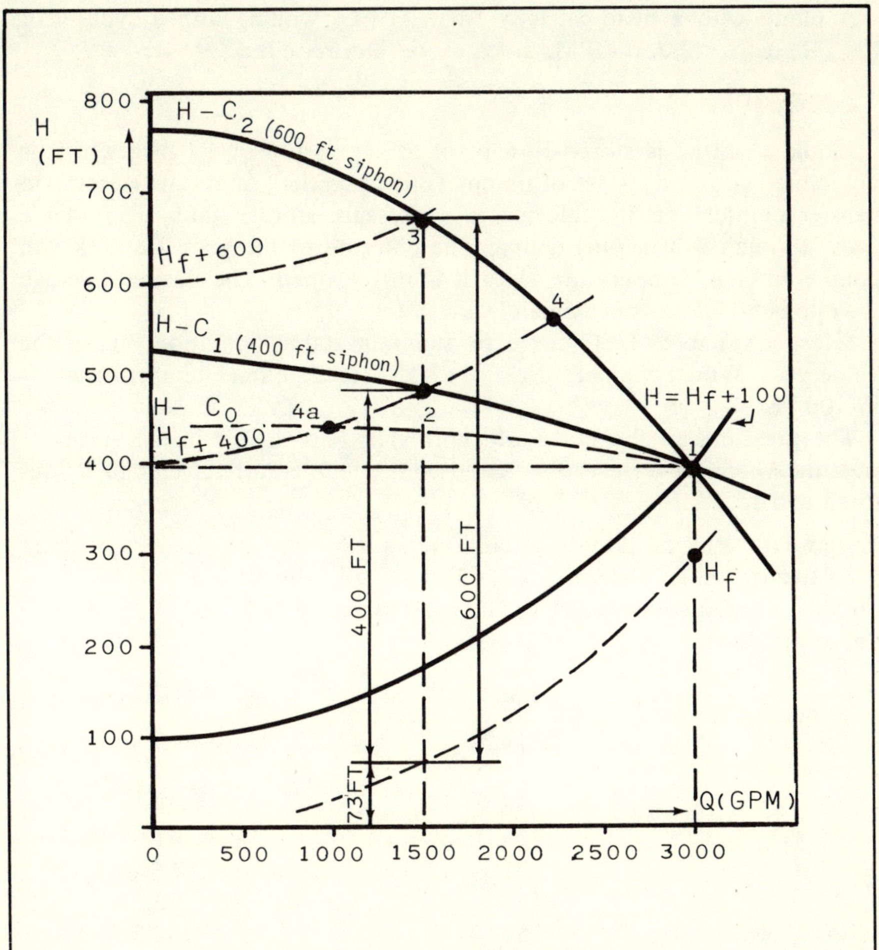

FIG. 15.12
Selection of head-capacity curves for syphon application.

Comments:

Point 1 = operating point = 3,000 GPM at 394 FT (294 friction loss);

$H-C_1$ = pump head-capacity curve to satisfy operating point as well as syphon requirement (6 fps, 1,500 GPM.).

Point 2 = operating point at start up = 1,500 GPM at 473 FT (friction loss = 73 FT, and static head = 400 FT);

$H-C_2$ = pump head capacity curve to satisfy operating point as well as syphon requirement.

Point 3 = operating point at start-up = 1,500 GPM at 673 FT (600 FT static head at start-up [syphon] and 73 FT loss as before).

A pump with a head capacity-curve $H-C_0$ would, with a 400-ft syphon, deliver only 950 GPM (point 4a), less than required for the case.

A pump with a head-capacity curve $H-C_2$ would, with a syphon of 400 FT, deliver 2,230 GPM, much more than required (pt. 4).

■ EXAMPLE 15:9

Liquid chlorine is stored in a plant at a pressure of 17 psig which is maintained constant. A set of pumps (one operating, one spare) serves to transfer uninterruptedly chlorine to a pressure storage tank, 15,000 FT away, through a 3-in buried line. The chlorine in the pressure tank will assume ambient temperature since it is in the open. The desired average flowrate is 50 GPM approximately.

Select a suitable $H-C$ curve to maintain stable operation during the whole year. Winter temperatures = 15-55°, and summer temperatures = 55-100° F.

The pressures in the tanks will correspond to the saturation (vapor) pressure which, in turn, is a function of the temperature. Refer to figures 15.13 and 15.14.

Design data: specific gravity of chlorine = 1.5, average viscosity = 0.25 CS; capacity (Q), average = 50 GPM, length plus fittings = 15,000 + 1,000 = 16,000 equivalent FT, and pipe diameter = 3″.
Vapor pressures:

° F	PSIA	PSIG
− 29	* 14.7	0
+ 5	31.7	17
+ 15	38.7	24
+ 35	56.0	41.3
+ 55	78.7	64
+ 75	108	93.3
+100	157	142.3

*atmospheric.
1 psi = 2.31/1.5 = 1.54 FT of chlorine.

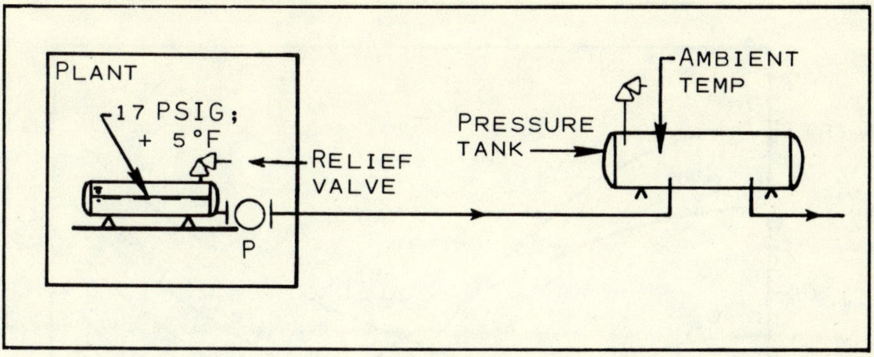

FIG. 15.13

The minimum and maximum pressure differentials between tanks are:

Winter:

min	= 24-17	= 7 PSI	= 11 FT of chlorine (15° F)
max	= 64-17	= 47 PSI	= 72 FT of chlorine (55° F)
average	= 41-17	= 24 PSI	= 37 FT of chlorine (35° F)

Summer:

min	= 64-17	= 47 PSI	= 72 FT of chlorine (55° F)
max	= 142.3-17	= 125.3 PSI	= 193 FT of chlorine (100° F)
average	= 93.3-17	= 76.3 PSI	= 118 FT of chlorine (75° F)

Compute the friction system head relationship:

$$H_f = f\frac{Le}{D^5} \times \frac{Q^2}{32.18}$$

$RE = 3162 \times Q_{AVERAGE}/(D \times Cs) = 3162 \times 50/(3 \times 0.25) = 210{,}800$

$f = 0.018$ (from charts)

$H_f = 0.018 \times 16{,}000 \times Q^2/(3^5 \times 32.18) = Q^2/27.2$ FT

Points for $H_f = Q/27.2$:

Q (GPM)	15	30	50	65	80
H_f (FT)	8	33	92	155	235

Construct H_f and H curves and superimpose H−C curves. A variety of pump curves could be selected for the case. The two curves shown should not necessarily be considered the optimum solution. They just serve the purpose of illustrating the criteria the design engineer must consider.

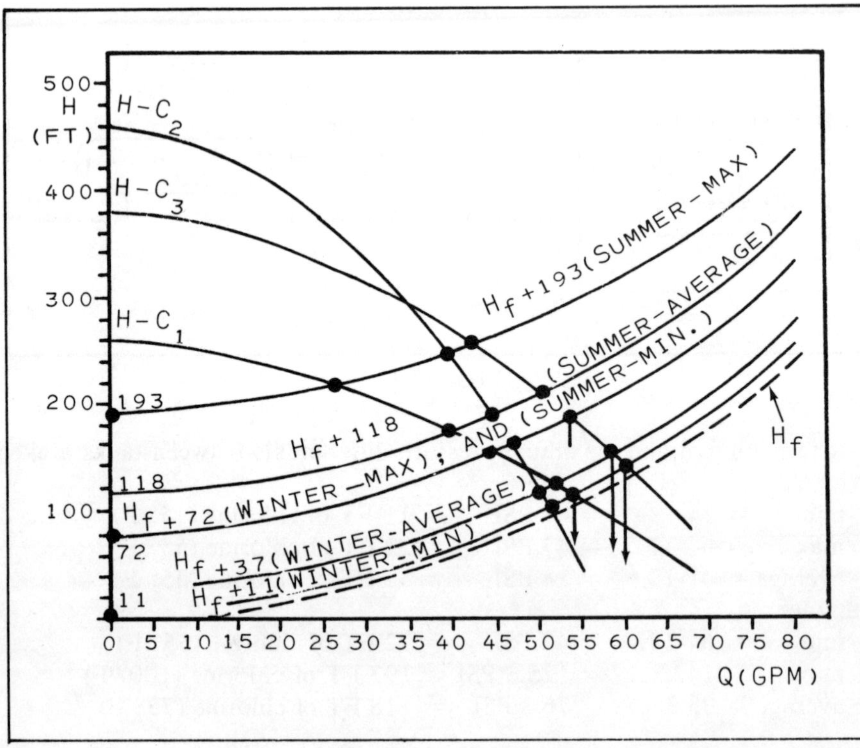

FIG. 15.14
Pump head-capacity curves to satisfy variable pressure differentials.

Capacities, GPM (refer to fig. 15.14)

	$H-C_1$	$H-C_2$	$H-C_3$
Winter min.	54	51	60
Winter avg.	52	50	58
Winter max.	45	47	53
Summer min.	45	47	53
Summer avg.	40	45	50
Summer max.	27	40	43

Comments: A steep curve like $H-C_2$ would prove to be a better selection. Comparing the capacities above, one can see that the $H-C_2$ will provide better results, especially in winter. $H-C_3$ will provide a very good summer average, but will exceed the demand during the winter. Depending on the requirements and the process, the design engineer can make an efficient selection with respect to capacities, heads, horsepower, and changes in the future.

■ EXAMPLE 15:10

A boiler feed pump is required to operate stably for capacities between 40 and 100 GPM. The boiler requirements are controlled by a feedwater regulator which actuates a throttling valve. Select a suitable pump head-capacity curve for a boiler generating steam at a pressure of 15 PSIG. Refer to Fig. 15.15; f=0.025, and sp.gr.=0.98 (water at 4 PSIA).

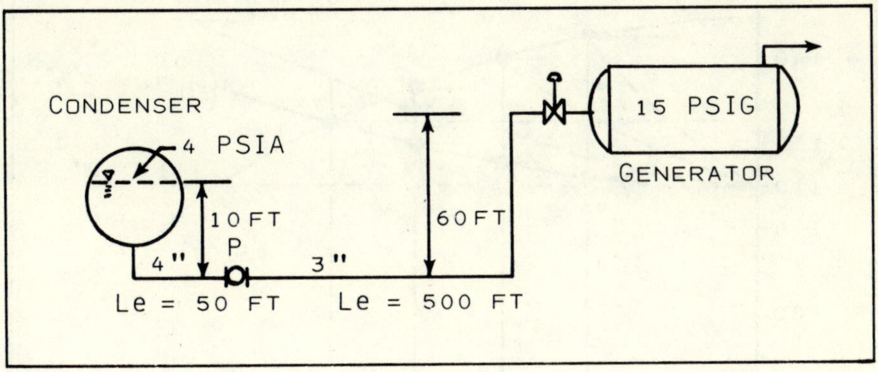

FIG. 15.15

For the calculations, $P_1 = 4\times 2.31/0.98 = 9.42$ FT, $P_2 = (15+14.7)\times 2.31/0.98=70$ FT, $Z_1=10$ FT, and $Z_2=60$ FT.

$$H_f=H_s+H_d$$

$$H_s=0.025\frac{50}{4^5}\times\frac{100^2}{32.18}= 0.4 \text{ FT}$$

$$H_d=0.025\frac{500}{3^5}\times\frac{100^2}{32.18}=16 \text{ FT}$$

$$H_f=0.4+16=16.4 \text{ FT}$$

$$H_f=\frac{0.025}{32.18}\times Q^2\left[\frac{50}{4^5}+\frac{500}{3^5}\right]=Q^2/611.$$

$$P_1+Z_1-H_f+E_p=P_2+Z_2$$

$$V_1=V_2=0$$

$$E_p=P_2+Z_2-P_1-Z_1+H_f=70+60-9.42-10+H_f=H_f+110.58$$

Maximum operating point is 100 GPM at 127 FT. (110.58—differential head; 16.4—friction loss). Construct H_f and H curves, superimpose a suitable H–C curve and comment.

Points for the H_f curve, fig. 15.16:

Q(GPM)	50	75	100	125
H_f(FT)	4	9.2	16.4	26

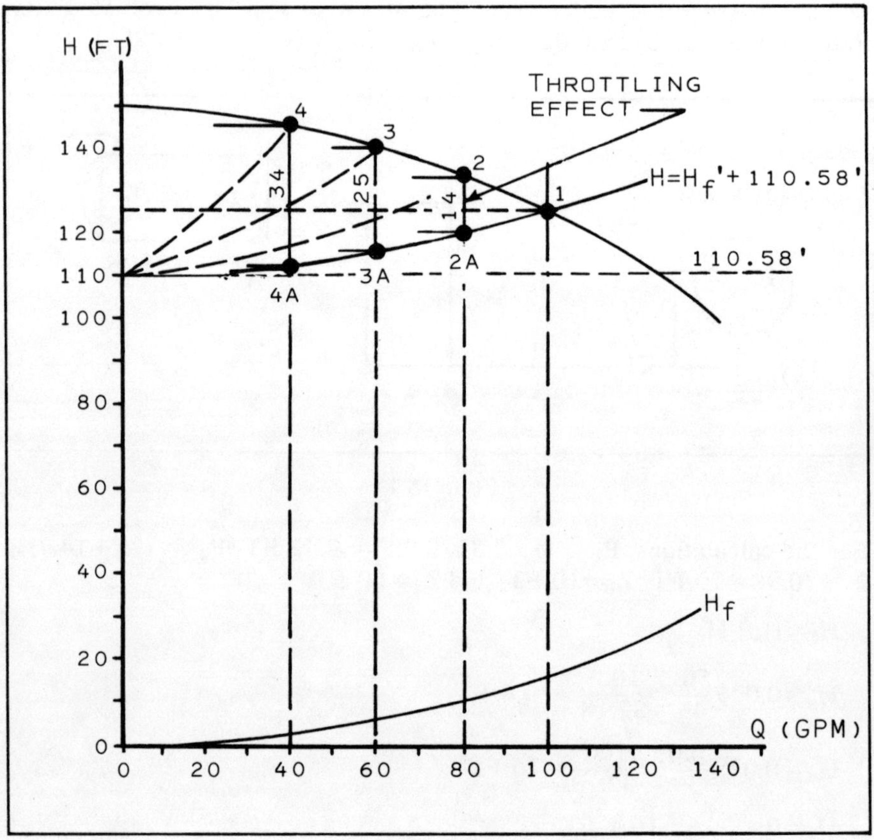

FIG. 15.16
Graphical presentation of throttling.

Refer to Fig. 15.16:

Point 1 = rated capacity = 100 GPM at 127 FT of condensate; throttling valve fully open.

Point 2 = Operating point = 80 GPM at 134 FT; throttling valve partially closed.

Point 2a = System head at this point = 120 FT. Head throttled in valve = 134 − 120 = 14 FT.

Point 3 = Operating point = 60 GPM at 141 FT; throttling valve partially closed.

Point 3a = System head at this point = 116 FT. Head throttled = 141 − 116 = 25 FT.

Point 4 = Operating point = 40 GPM at 147 FT; throttling valve partially closed.

Point 4a = System head at this point = 113 FT. Head throttled = 147 − 113 = 34 FT.

Comments:

It should be borne in mind that the only H-curve is 110–1. The literature indicates system curves 110–2; 110–3; 110–4, although these are nonexistent. The system stays constant, and nothing changes in it (same diameters, same lengths). Therefore, H_f remains the same function of Q.

The only correct way to indicate variable capacities, achieved by means of throttling, is per fig. 15.16, omitting the broken lines. In case point 3 is the operating point, pump head = 141 FT of which 110.58 FT represent the static head, 25 FT the head loss in the valve, and 5.42 FT the friction loss in the system.

■ EXAMPLE 15.11

Discuss a pumping system involving two or more discharge branch lines for simultaneous transfer to two or more locations.

1. BRANCH-OFF AT PUMP DISCHARGE

This is the case when the line before the branch-off, and the suction line are very short, and their losses are neglected.

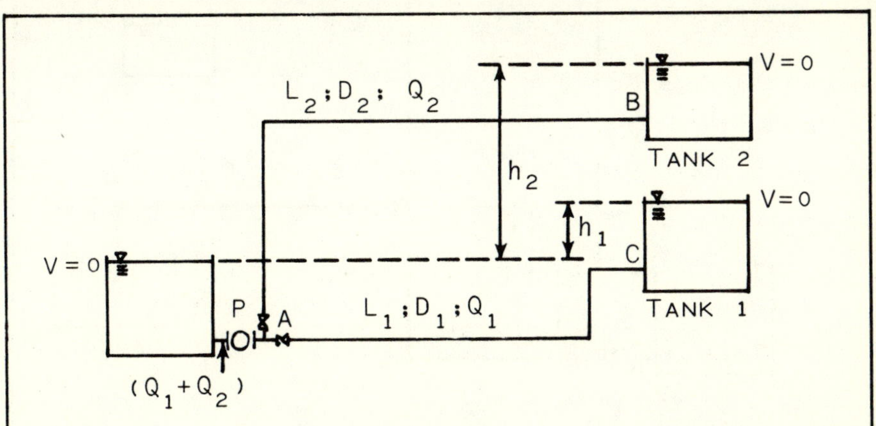

FIG. 15.17

In the arrangement per fig. 15.17, once flow to the two consumers has been established, the system head for branch AB will be equivalent to the system head for branch AC. Flow will reach tank 2 only when the friction head loss from A to C will exceed the static elevation ($h_2 - h_1$). (Refer to Chapter 6).

$$\text{System head for AB} = H_{AB} = f_2 \frac{L_2}{D_2^5} \times \frac{Q_2^2}{32.18} + h_2$$

$$\text{System head for AC} = H_{AC} = f_1 \frac{L_1}{D_1^5} \times \frac{Q_1^2}{32.18} + h_1$$

For assumed $f = f_{AB} = f_{AC}$, and $H_{AB} = H_{AC}$:

$$\frac{f \times L_2 \times Q_2^2}{D_2^5 \times 32.18} + h_2 = \frac{f \times L_1 \times Q_1^2}{D_1^5 \times 32.18} + h_1$$

Types of problems with this arrangement:

1.1 Given L_1, L_2, D_1, D_2, $(Q_1 + Q_2)$, elevations. Select H–C curve and determine Q_1 and Q_2 and the total head.

1.2 Given Q_1, Q_2, L_1, L_2, D_1, D_2, h_1, h_2. Select H–C curve and determine the total head.

2. BRANCH-OFF AT A DISTANCE FROM PUMP.

In this case, there is a common line of appreciable length.

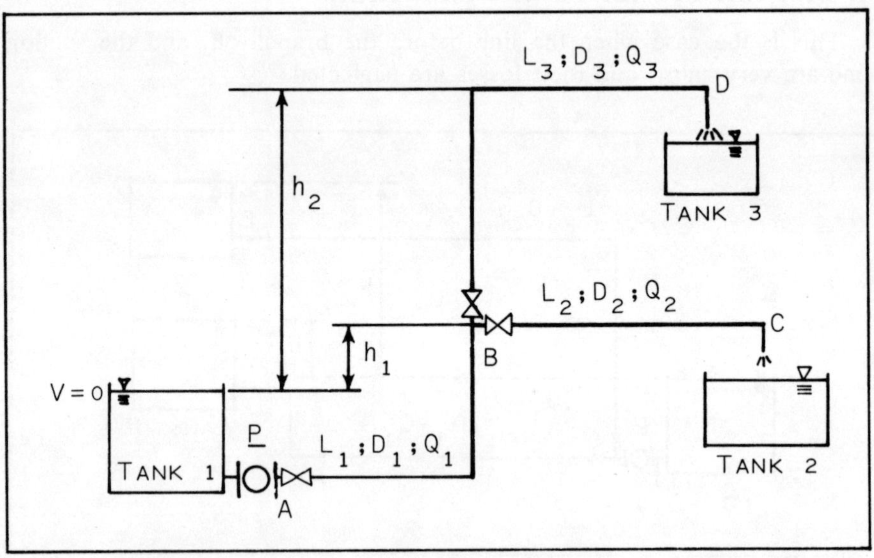

FIG. 15.18

In FIG. 15.18, AB represents the common line, and $Q_1 = Q_2 + Q_3$. Per chapter 6, the hydraulic gradient B′ for junction point B is equal for branch BD and branch BC. It follows that the system heads for these branches will be equivalent. Flow will reach tank 3 only when the friction losses in branch BC exceed $(h_2 - h_1)$.

Types of problems with this arrangement are:

2.1. Given all L's, D's, Q_1, and static elevations, select H–C curve and determine Q_2, Q_3, and the total head.

2.2. Given Q_2 and Q_3, all L's and D's, and static elevations, select H–C curve and determine the total head.

■ EXAMPLE 15:12

Refer to fig. 15.17.

Data: $L_1, L_2 = 2{,}000$ FT, 1,500 FT; $D_1, D_2 = 6'', 4''$; $h_1, h_2 = 20$ FT, 40 FT; $Q_1 + Q_2 = 1{,}200$ GPM; and $f_1 = f_2 = 0.016$.

$$H_{f1} = 0.016\frac{2000}{6^5}\times\frac{Q^2}{32.18} = Q^2/7820$$

$$H_1 = 20 + Q^2/7820$$

$$H_{f2} = 0.016\frac{1500}{4^5}\times\frac{Q^2}{32.18} = Q^2/1373$$

$$H_2 = 40 + Q^2/1373$$

Plot the individual system-head curves, and thereafter the combined system-head curve (add capacities at equal heads, operation in parallel); superimpose a suitable H–C curve for total flowrate. Read off the total head as well as operating points for individual transfers to tanks.

Points for the System-Head Curves:

Q(GPM)	200	500	800	1,000
H_1(FT)	25	52	102	148

Q(GPM)	100	200	400
H_2(FT)	47	69	156

Refer to FIG 15.19:

S − T − S′ = Combined System Curve (S–2 + N–3 at same head)
H_1 = S − Tank 1(S–2); H_2 = N − Tank 2(N–3).

Point 1 = operating point for simultaneous flow: 1,200 GPM at 118 FT.

T = for flowrates and heads less than 400 GPM and 40 FT, no flow

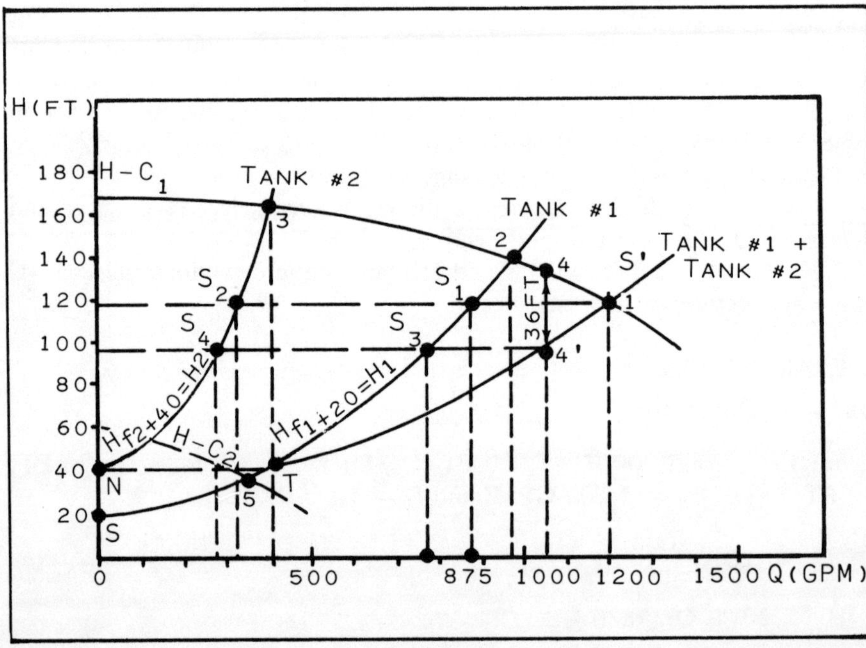

FIG. 15.19

will reach tank 2 (static head $h_2 - h_1 = 20$ FT, and friction loss of 20 FT for Q = 400 GPM in AC).

S_1 = Transfer to tank 1 at the rate of ∿ 875 GPM. Friction loss in branch = 118 − 20 = 98 FT.

S_2 = Transfer to tank 2 at the rate of ∿ 325 GPM. Friction loss in branch = 118 − 40 = 78 FT. (875 + 325 = 1,200).

If the valve at the pump discharge is throttled to absorb 36 FT (4–4′), the pump will operate at point 4: 1,050 GPM at 132 FT, with delivery rates to tank 1 and 2 = 775 GPM (S_3), and 275 GPM (S_4). System heads for S_3 and S_4 = 96 FT.

Point 2 = individual flowrate to tank 1 (no delivery to tank 2): 970 GPM at 140 FT (friction loss = 140–20 = 120 FT)

Point 3 = individual flowrate to tank 2 (no delivery to tank 1): 409 GPM at 162 FT (friction loss = 162–40 = 122 FT)

Note: A pump with the characteristic head-capacity curve $H–C_2$ would operate at point 5: 350 GPM at 35 FT, and deliver only to tank 1. Delivery to both tanks starts at point T which it can not reach.

■ EXAMPLE 15:13

Refer to fig. 15.17. The required branch flowrates, Q_1 = 1000 GPM, and

Q_2 = 450 GPM. Branch diameters, $D_1 = 6''$, and $D_2 = 4''$. Static heads h_1 = 10 FT and h_2 = 20 FT. Branch equivalent lengths, L_1 = 1,000 FT, and L_2 = 500 FT. Assume f for both branches to equal 0.016.

Comment: The above data have to satisfy the condition that the system heads for the branches have to be equal. If they are not, there will be a deviation from the capacities above: If we decide that Q_1 = 1,000 GPM has to stay, then we compute Q_2:

$$H_1 = 0.016\frac{1000}{6^5}\times\frac{1000^2}{32.18} + 10 = 74 \text{ FT}$$

$$H_2 = 74 = 0.016\times\frac{500}{4^5}\times\frac{Q_2^2}{32.18} + 20$$

$$Q_2 = 472 \text{ GPM}$$

Thus the corrected Q_2 will amount to 472 GPM. If the process requires 450 GPM, a control valve will be installed to regulate this flowrate, but, for design purposes, 472 GPM will have to be used.

Thus, the pump operating point is = 1,472 GPM at 74 FT.

$$H_1 = 0.016\times\frac{1000}{6^5}\times\frac{Q^2}{32.18} + 10 = Q^2/15{,}640 + 10$$

$$H_2 = 0.016\times\frac{500}{4^5}\times\frac{Q^2}{32.18} + 20 = Q^2/4{,}120 + 20$$

Data for the system-head curves:

Q(GPM)	300	600	1,000	1,300
H_1(FT)	16	33	74	118
Q(GPM)	100	200	400	600
H_2(FT)	23	30	59	107

Refer to FIG. 15.20.

S – T – S′ = the combined system curve (S–2 + N–3 at the same head).

Point 1 = operating point = 1,472 GPM at 74 FT (simultaneous delivery).

T = for flowrates and head lower than 375 GPM and ∿ 20 FT no flow will reach tank 2. (10 FT static head and ∿ 10 FT friction loss in AC)

S_1 = delivery to tank 1: 1,000 GPM (friction loss in branch: 74–10 = 64 FT).

S_2 = delivery to tank 2: 472 GPM (friction loss: 74–20 = 54 FT).

Point 2 = individual flowrate to tank 1 : 1,085 GPM at 85 FT (friction loss: 75 FT; no flow to tank 2).

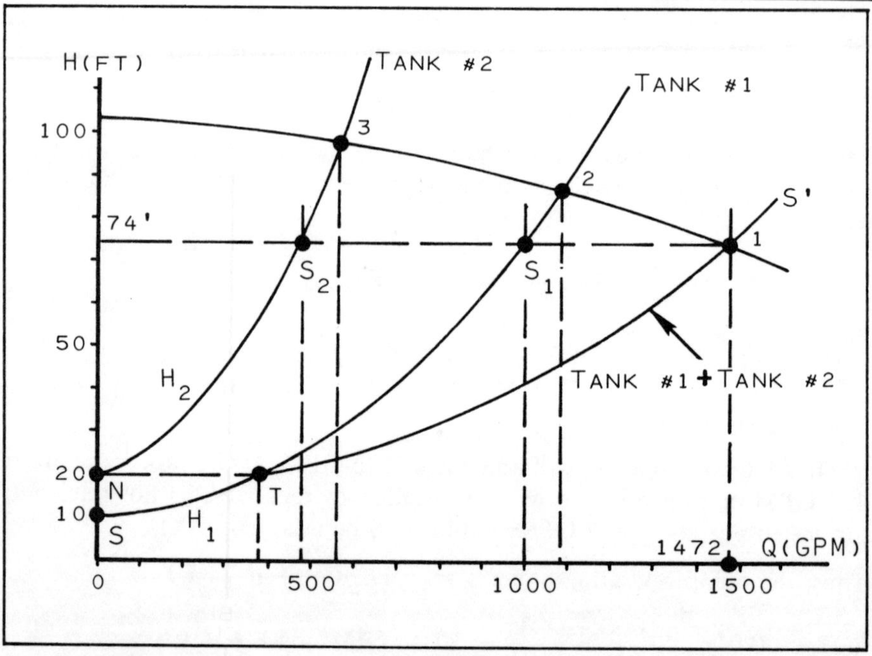

FIG. 15.20

Point 3 = individual flowrate to tank 2 : 560 GPM at 96 FT (friction loss 76 FT; no flow to tank 1)

- EXAMPLE 15:14 (Refer to fig. 15.18)

Select a suitable head-capacity curve to satisfy the following conditions:
Q_1 = 4,000 GPM; L_1, L_2, L_3 = 800 FT, 1,000 FT, 900 FT; D_1, D_2, D_3 = 12″, 8″, 6″; h_1, h_2 = 30 FT, 60 FT, and f_1, f_2, f_3 = 0.02.
Determine the individual branch capacities.

Solution Procedure:

1. Compute and plot friction loss system curves AB, BC, BD.
2. With reference to branch point B, plot branch system head-capacity curves: BC, and (BD + h_2 − h_1) = (BD + 60 − 30) = (BD + 30)
3. Plot system-head curve (AB + h_1) = (AB + 30).
4. Add (AB + 30) and (BD + 30) (heads for equal capacities), and obtain [(BD + 30) + (AB + 30)]. This curve is necessary for the case in which the pump will supply product to tank 3 only.
5. Add (AB + 30) to BC (heads for equal capacities), and obtain [BC + (AB + 30)]. This curve is necessary for the case in which the pump will supply product to tank 2 only.
6. Obtain the common branch system curve [(BD + 30) + BC] by adding curves (BD + 30) and BC (add capacities for equal heads).

7. Add to curve [(BD + 30) + BC] the curve (AB + 30) (heads for equal capacities), and obtain the total system-head curve, running from head 30 FT to point 1. Refer to fig. 15.21.

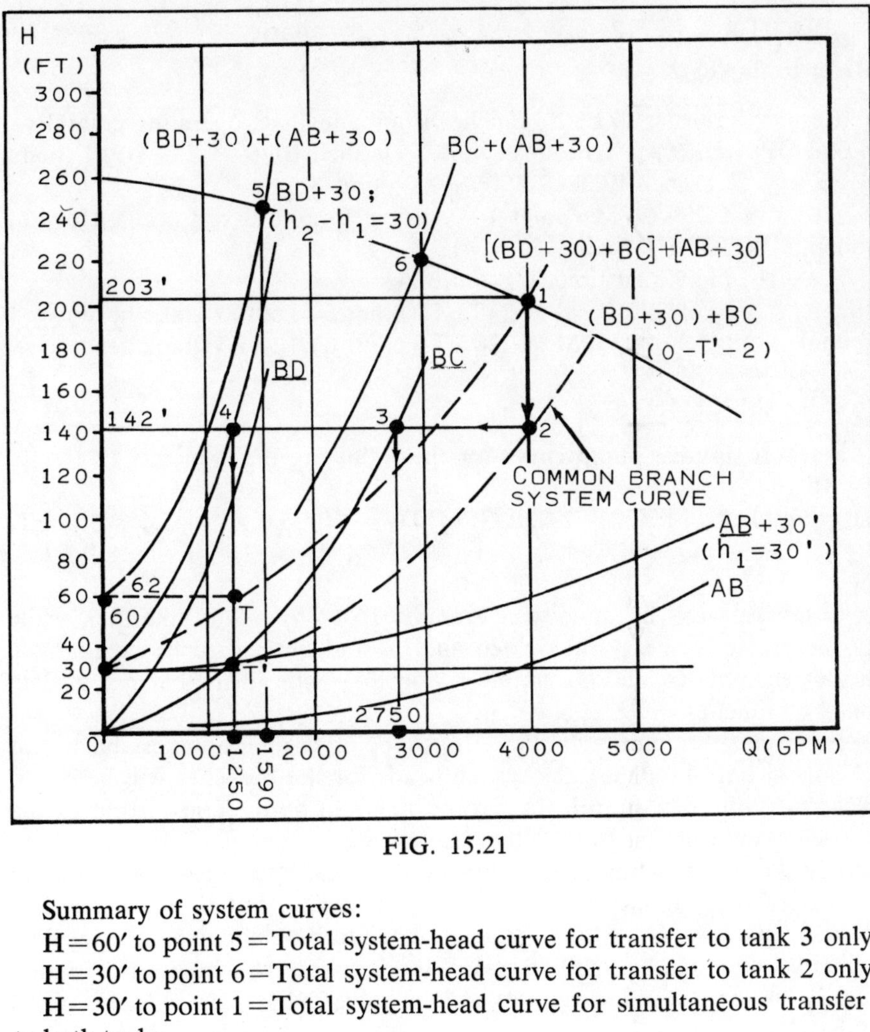

FIG. 15.21

Summary of system curves:

H=60′ to point 5=Total system-head curve for transfer to tank 3 only.

H=30′ to point 6=Total system-head curve for transfer to tank 2 only.

H=30′ to point 1=Total system-head curve for simultaneous transfer to both tanks.

$$AB = 0.02\frac{800 \times Q^2}{12^5 \times 32.18} = Q^2/500{,}500$$

$$BC = 0.02\frac{1000 \times Q^2}{8^5 \times 32.18} = Q^2/53{,}000$$

$$BD = 0.02\frac{900 \times Q^2}{6^5 \times 32.18} = Q^2/13{,}900$$

Points for the system head curves:

Q(GPM)	1,000	1,500	2,000	2,500	3,000	5,000
AB(FT)	2		8		18	50
BC(FT)	19	43	76	118	170	
BD(FT)	72	161	287	450		

Refer to fig. 15.21.

Results: Transfer to both branches simultaneously: operating point 1 = 4,000 GPM at 203 FT head: to tank 3 (branch BD)—1,250 GPM (point 4) to tank 2 (branch BC)—2,750 GPM (point 3).

At up to 1,250 GPM (point T), no flow to tank 3 will occur. Head at junction B: 142 FT (points 2, 3, 4).

Transfer to individual branches and tanks:

to tank 3 only: 1,590 GPM at 247 FT (point 5) (60 FT static head)
to tank 2 only: 3,000 GPM at 220 FT (point 6) (30 FT static head)

■ EXAMPLE 15.15

Select a suitable pump curve for the following requirements (Refer to fig 15.18):

$Q_2 = 4{,}000$ GPM; $Q_3 = 2{,}000$ GPM; $D_2 = 10''$, $D_3 = 8''$, $L_2 = 365$ FT, $L_3 = 500$ FT; $Q_1 = Q_2 + Q_3 = 6{,}000$ GPM; $D_1 = 16''$; $L_1 = 300$ FT, $h_1 = 30$ FT; $h_2 = 50$ FT; and $f = 0.02$.

Comparing this example with example 15.14, one will note that, while in the previous one Q_1 was given and Q_2 and Q_3 had to be determined, in this example Q_2 and Q_3 are given, together with all the balance of data on the branches.

As we determined previously, in this type of system, upon establishment of flow in both branches, the system heads for the branches will be equal. Therefore, the first step in this problem should be the comparison of the branch system heads. In case they are not equal, one wouldn't be able to satisfy the predetermined conditions, unless one introduces some change.

Any of the parameters can be changed to suit, but the easiest would be the capacities. If all the parameters have to be maintained as specified, an adjustment can be made in one of the branches to increase the friction loss by means of the inclusion of a globe valve.

For this case:

$$H_{BC} = 0.02\frac{365}{10^5}\times\frac{4000^2}{32.18} = 36.5 \text{ FT}$$

$$H_{BD} = (50 - 30) + 0.02\frac{500}{8^5}\times\frac{2000^2}{32.18} = 58 \text{ FT}$$

$H_{BC} \neq H_{BD}$.

A globe valve will have to be included in BC. Tables indicate that the "K" factor for a 10″ globe valve equals 5. Therefore, the friction loss through it, when open, would be $5 \times V^2/2g$. The velocity head for a 10″ line and 4,000 GPM equals 4.12 FT. Thus, the friction loss will be equal to $5 \times 4.12 = 21.5$ FT. The addition of this loss would establish system head equivalency and allow the flowrates required.

The energy wasted would be equivalent to approximately $4{,}000 \times 21.5/(3{,}960 \times 0.75) = 30$ HP. Depending on the frequency of operation, this loss can be substantial or negligible. The equivalent length for a 10″ globe valve is 215 FT. Thus the branch system head equations would read:

$$H_{BC}=0.02\frac{365+215}{10^5}\times\frac{Q^2}{32.18}=Q^2/277{,}400$$

$$H_{BD}=20+0.02\frac{500}{8^5}\times\frac{Q^2}{32.18}=20+Q^2/105{,}450$$

Upon computation of the main line system head curve, the procedure is identical as in example 15.14.

▪ EXAMPLE 15.16

Select a suitable pump for a drydock to operate stably at capacities and heads ranging from 50,000-100,000 GPM and 15-50 FT of liquid respectively.

Mixed flow pumps are suitable for operations within these ranges. (Refer to chapter 9, item 9.2, and chapter 12). They are also used for irrigation, drainage, storm water, and circulating systems.

A ship repair drydock is a shoebox type structure at a river or sea, with its highest point at river or sea-water level. One of its sides is removable so water and ship can enter the dock. The pumps then empty the drydock to permit work around the ship.

In this case, the pump suction head is variable, and the discharge head is constant. Refer to Fig. 15.22.

The pumps are installed below the bottom of the drydock. Thus the NPSHA and submergence requirements are ample. Very large discharge ducts are used, thus discharge friction losses are very small. In order to save time, several pumps are installed; although they operate simultaneously, they are independent of each other. Refer to schematic arrangement per fig. 15.22.

CASE 1. Dock full; Total head $= H_{max} - H_{max} + H_f = H_f$

CASE 2. Water level in dock$=h_{min}$

Total head$=H_{max} - h_{min} + H_f = 45 + H_f$

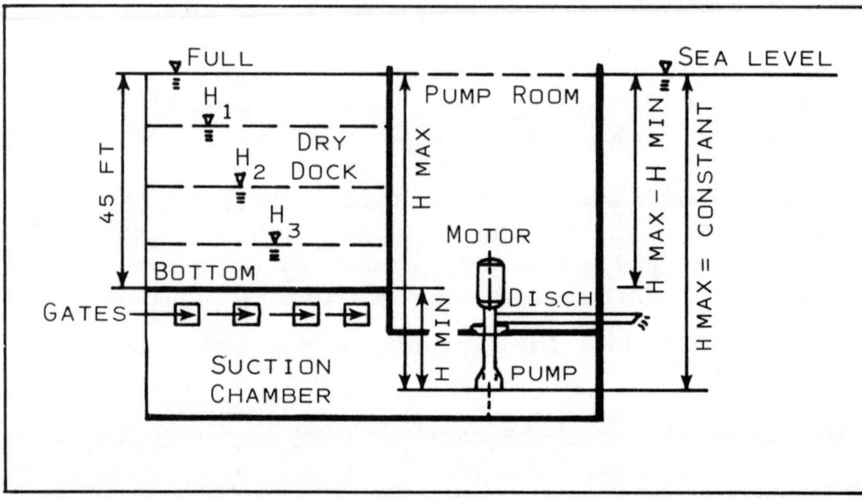

FIG. 15.22

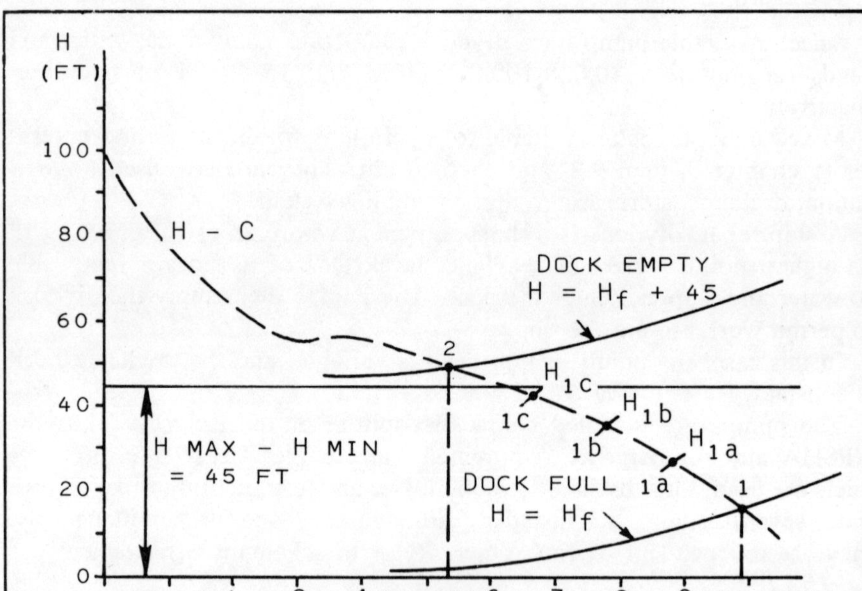

FIG. 15.23
Mixed flow pump curve for drydock application.

Refer to Fig 15.23.

Note: Mixed flow pumps are rarely operated below capacities for which the power requirements rise sharply, approximately 40,000 GPM and lower on FIG 15.23. These pumps are not started at shut-off as volute centrifugal pumps are. Refer to Chapter 12.

In figure 15.23, Point 1 = operating point at start of emptying operation (dock full)—appr. 100,000 GPM at 15 FT, Point 2 = end of operation (dock empty)—53,330 GPM at 50 FT. H_{1a}, H_{1b}, and H_{1c} = portions of intermediate system-head curves and 1_a, 1_b, and 1_c = intermediate points of operation.

16.

Application of Centrifugal Pumps (Multiple Pump Operations for New Systems)

16.1 PUMPS IN PARALLEL

Generally, pumps are arranged to operate in parallel to increase the capacity of a system, provide flexibility with respect to capacity, and to some extent, head; and, in case of emergency, provide the services of a spare pump.

Most pumping installations require a spare pump. When a single pump is capable of providing the required service, the spare will be an identical unit. This system is called, in short, "one pump operating, one spare". In reality, the function of operating and spare in this case alternates between the two pumps on a weekly or monthly basis.

When deciding on multiple pumps, the application engineer must analyze carefully whether to recommend one pump with one spare, two pumps in parallel with one pump spare, two pumps in parallel without a 100% spare (one pump runs by itself in case the other has to be cut out of service, thus the capacity will be 50% of the total), provided that the process can tolerate less capacity at times; three pumps, four etc.

The number of combinations is large; so is the number of pump manufacturers. Some manufacturers may not be able to offer a single pump for the capacity and head required, so they offer two pumps in parallel. Other manufacturers can offer a single pump for this service.

An evaluation is mandatory. Various factors have to be investigated: cost of two pumps versus three, space available, operation and maintenance, flexibility of operation, and variations required. A careful study is required to get to the optimum solution.

It is imperative that the application engineer be familiar with the whole process, the operating requirements and procedures.

A pump selection based solely on the requirement to provide Q GPM at H FT will definitely turn out to be a costly venture. Pumps do not have necessarily to be identical when used in parallel. Sometimes, because of lack of knowledge about the process, there is the tendency to select identical pumps, instead of investigating the operation in parallel of two pumps with dissimilar curves which could prove to be a more efficient and versatile operation. (Refer to chapter 14)

■ EXAMPLE 16.1

Select suitable pump head-capacity curves to provide efficiently a varying water supply, ranging from 2,000 to 4,000 GPM. Pump operation will be intermittent, and minimum capacity during emergency (one pump idle) shall be considered adequate. Comment on the efficiency of the operation upon selection of the pump curves. Refer to fig. 16.01.

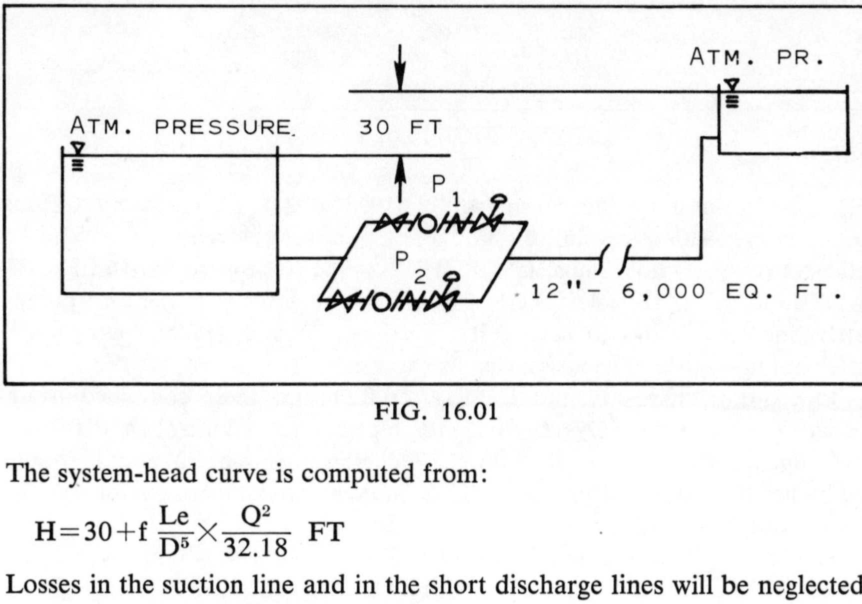

FIG. 16.01

The system-head curve is computed from:

$$H = 30 + f\,\frac{Le}{D^5} \times \frac{Q^2}{32.18}\ \text{FT}$$

Losses in the suction line and in the short discharge lines will be neglected.

$$RE = \frac{3{,}162 \times 4{,}000}{12 \times 1} = 1{,}054{,}000$$

f=0.015 (New Line)

$$H = 30 + 0.015\frac{6000}{12^5} \times \frac{Q^2}{32.18} = 30 + Q^2/89{,}000$$

Points for the system-head curve:

Q (GPM)	500	1,000	2,000	3,000	4,000
H (FT)	33	41	75	131	210

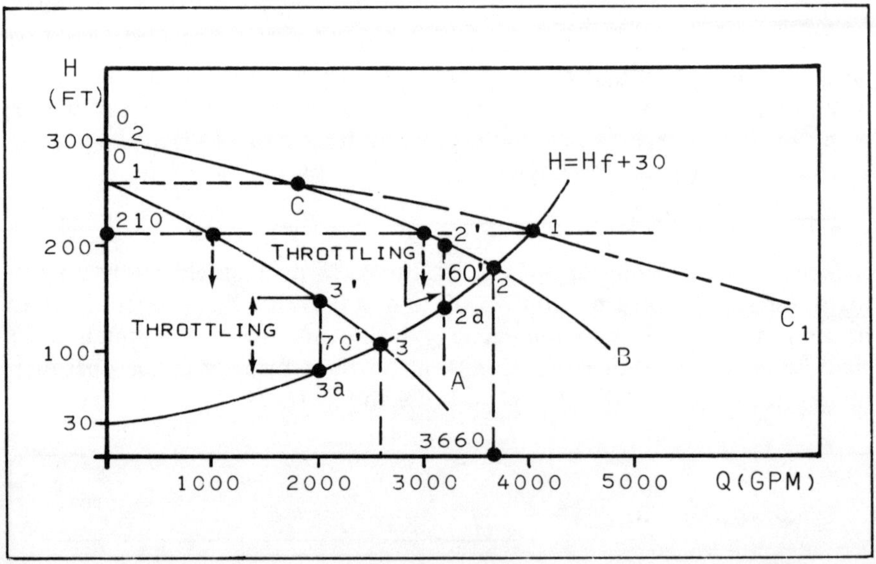

FIG. 16.02

Establish point 1 on the curve (4,000 GPM at 210 FT) and try various pump curves (identical and dissimilar). Operation, as required, should be efficient for the whole range (2,000-4,000 GPM). Refer to fig. 16.02.

Comments: 2,000 − 4,000 GPM is a wide range for two identical pumps operating in parallel to satisfy it efficiently. Two dissimilar pumps will offer a much more efficient service as shown.

Plot pump curves A and B (Fig. 16.02) and their combined head-capacity curve A+B (O_2-C-C_1). Each pump has a control valve to control capacity by means of throttling. Efficiencies at points 2′ and 3′ should be higher than those at points 2 and 3, in order to compensate for the lost energy during throttling.

1. Main operating point is 4,000 GPM at 210 FT. Pump A delivers 1,000 GPM at 210 FT., and pump B, 3,000 GPM at 210 FT.

2. Pump B (O_2-C-B) by itself delivers 3,660 GPM at 180 FT. When throttling, the pump can operate at 2′ (3,200 GPM at 200 FT), and 60 FT are lost in the valve (2a).

3. Pump A (O_1-A) by itself delivers 2,600 GPM at 106 FT. With throttling (point 3′) the pump will transfer 2,000 GPM at 150 ft. Loss in the valve is 70 FT.

Summary of capacities which can be provided efficiently: 2,000 GPM (pump A); 2,600 GPM (pump A); 3,200 GPM (pump B); 3,660 GPM (pump B); 4,000 GPM (pump A+B).

■ EXAMPLE 16.2

Providing varying capacities to a consumer is achieved by having a number of pumps, arranged for parallel operation, to operate singly or together. As shown in the previous example, additional flexibility can be added by means of control valves.

When the pumps, arranged in parallel, are close to each other, the losses between the pump discharge and junction point, from which the common discharge line starts, are usually neglected because they are small when compared to overall losses. Not so, when the branch lines are long. The losses in them can be appreciable, and have to be taken into consideration.

In fig. 16.03 we have an arrangement in which two identical pumps are 1,000 equivalent FT away from the junction point. The branch lines are 8″ each, and the common line is 12″. Neglect the losses from the suction tanks to the pumps, and select two identical pumps to provide a capacity of 3,800 GPM when running in parallel, and 2,500 GPM when only one pump will be operating.

The equivalent length of the common 12″ line is 4,000 FT. The equivalent length of the heat exchanger is 1,000 FT. Assume a friction factor of f=0.018 for all pipes. The liquid levels in all three tanks are 50 FT and constant. Pressure is atmospheric (14.7 PSIA).

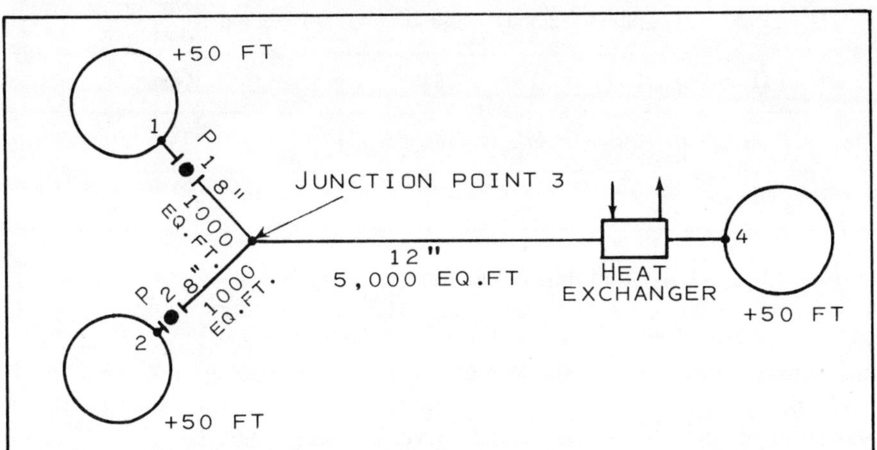

FIG. 16.03

For one pump operating singly:

$$H_{f1,3,4}\ (\text{or } H_{f2,3,4}) = f\frac{L_{1,3}}{D_{1,3}{}^5}\times\frac{Q^2}{32.18}+f\frac{L_{3,4}}{D_{3,4}{}^5}\times\frac{Q^2}{32.18}$$

$$=\frac{0.018\times Q^2}{32.18}\left[\frac{1000}{8^5}+\frac{5000}{12^5}\right]=Q^2/35{,}325=H_{f2,3,4}$$

For two pumps operating in parallel, since branches 1-3 and 2-3 are equivalent, it follows that $H_{f1\text{-}3}$ and $H_{f2\text{-}3}$ are equivalent. This means that $Q_{1,3}=Q_{2,3}=0.5\times Q_{3,4}$ (since both pumps are identical, and taking suction from tanks at the same level).

$$H_{f3,4}=f\frac{L_{3,4}}{D_{3,4}{}^5}\times\frac{Q^2}{32.18}=0.018\frac{5{,}000}{12^5}\times\frac{Q^2}{32.18}=Q^2/88{,}970$$

The system curve for one branch, while two pumps are in operation, is:

$$H_{f1,3}=H_{f2,3}=f\frac{Le}{D^5}\times\frac{(0.5Q)^2}{32.18}=0.018\frac{1000}{8^5}\times\frac{0.25\times Q^2}{32.18}=Q^2/234{,}330$$

The total system curve for two pumps operating is:

$$H_{f3,4}+H_{f1,3}=Q^2\left[\frac{1}{88{,}970}+\frac{1}{234{,}330}\right]=Q^2/64{,}490$$

Points for the System-Head Curves:

Q (GPM)	1,000	2,000	3,000	4,000	5,000	No. of pumps operating
$H_{f1,3,4}$ (FT)	28	113	254			One
$H_{f3,4}$ + $H_{f1,3}$ (FT)	16	62	140	248	387	Two in parallel
$H_{f3,4}$ (FT)	11	45	101	180	281	System head curve for common line

Plot $H_{f1,3,4}$ $H_{f1,3}$ and $H_{f3,4}$ and superimpose a curve representing the combined curve for two pumps (fig. 16.04) Determine point 2 by requirement: the intersection of 3,800 GPM with $H_{f3,4}+H_{f1,3}$. Next halve the capacities and obtain the H$-$C curve of the single pump. Compare it with manufacturers' curves for similar curves. If unavailable, try a variation of the selected combined curve, crossing point 2.

For one pump operating, use system curve OA, point 1–2,450 GPM at 170 FT.

For two pumps in parallel, use system curve OB, point 2–3,800 GPM at 223 FT. The capacity of each pump is 1,900 GPM at 223 FT.

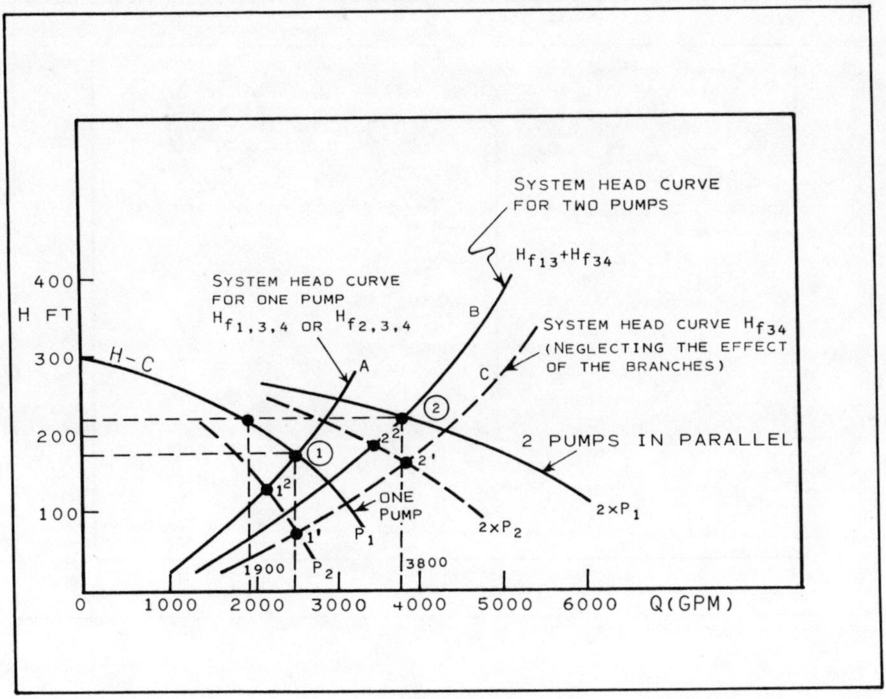

FIG. 16.04

Comment: If one neglects the branch losses, the system head curve will be represented by $H_{f3,4}$ = the system head curve for the common line only. The selected operating point would be then 2′ (3,800 GPM at 160 FT).

The combined pump curves, $2\times P_2$, would then result in the pump curve P_2 for the single pump. In this case, when one pump will be operating by itself, Point 1^2, instead of Point 1^1 will be the operating point (2,100 GPM, as compared to the 2,500 GPM required). The head at the junction point (parallel operation) = 160 FT (pt 2′) [From 223 FT $-(0.018\times 1{,}000/8^5)\times(1{,}900^2/32.18]$ The head at the junction point (single operation) = 67.5 FT (pt 1′) [From 170 FT $-(0.018\times 1000/8^5)\times(2{,}450^2/32.18)]$

■ EXAMPLE 16.3

Select two identical pumps to operate in parallel and transfer approximately 4,000 GPM of water at 60° F. Determine the heads at the pumps during operation in parallel.

As the arrangement in fig. 16.05 indicates, the two pumps are at a

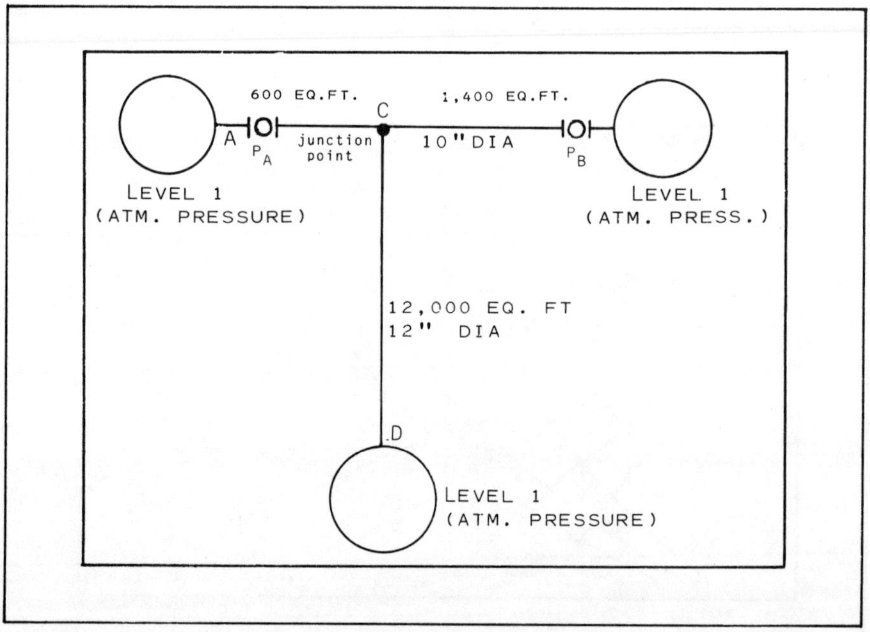

FIG. 16.05

different distance from the junction point. Determine also the capacities and heads for each pump operating singly. Determine the pressure at the junction point C for both types of operation.

The procedure is as follows: (Refer to fig. 16.06):

1. Since the pumps and their head-capacity curves will be identical, and since the pumps will be installed at a different distance from junction point C, they will have different operating points for an operation in parallel.

2. Compute and plot the following system head curves ($H=H_f$):

$H_{f\ AC}$ —from pump A to junction C.

$H_{f\ BC}$ —from pump B to junction C.

$H_{f\ CD}$ —from junction C to tank D.

$H_{f\ ACD}$—from pump A to C and D (single pump operation).

$H_{f\ BCD}$—from pump B to C and D (single pump operation).

3. The water level in the tanks is assumed equal and constant. The suction losses from the suction tanks to the pumps are included in the equivalent lengths indicated.

4. Draw a vertical line up from Q=4,000 GPM to its intersection with $H_{f\ CD}$, point 1 (Refer to fig. 16.06). This will be the operating point at junction C when both pumps will be operating in parallel.

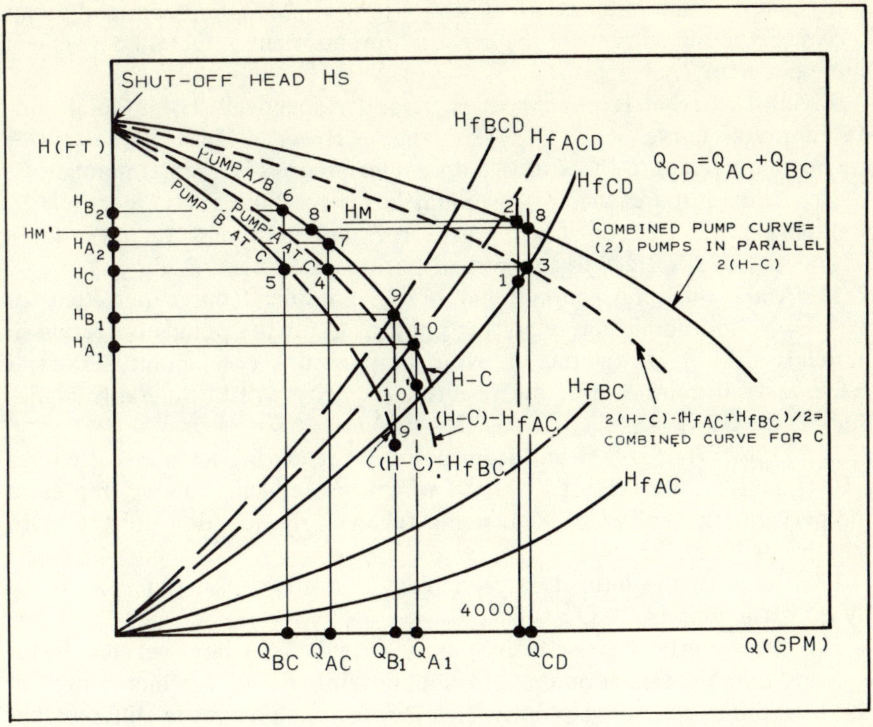

FIG. 16.06

5. Due to excessive distance between the pumps and junction point C, there will be a substantial loss of head from the pumps to the junction. The average loss in the branches during simultaneous flow is added above point 1, to increase the head for 4,000 GPM to H_M.

The average loss in the branches equals their sum divided by two. Note that, in the case of equidistant pumps from junction C, the average branch loss would equal the loss in one of the branches, provided their diameters are identical: $[(H_{fAB}+H_{fBC})/2=H_{fAC}=H_{fBC}]$.

6. Assume in this case the average loss to be equal to 30 FT. Add 30 FT to point 1 and obtain point 2. The combined pump curve for two pumps operating in parallel can be tentatively drawn through point 2 by feel. Obtain $2\times(H-C)$.

Upon completion of the construction of all curves, the true common point may shift from the assumed point 2. If the deviation is acceptable, the selected combined curve may be considered adequate. If not, a new assumption has to be made.

7. From $2\times(H-C)$ obtain the individual $H-C$ curve by halving capacities at equal heads.

8. Compare the obtained H−C curve to available manufacturers' pump curves. If too far out, repeat the construction per item 6. Obtain new H−C curve per item 7.

9. Subtract from H−C curves H_{fAC}, and respectively H_{fBC}. Obtain adjusted pump curves $(H-C)-H_{fAC}$, and $(H-C)-H_{fBC}$. These curves are now equivalent to those of two dissimilar pumps installed at junction C.

10. Add up the capacities for equal heads for both curves per item 9, and obtain the true combined curve for junction point C: $2(H-C)-(H_{fAC}+H_{fBC})/2$. (Refer to the second paragraph in item 5 above.)

11. Mark the intersection point of the combined curve per item 10 with $H_{f\ CD}$ and obtain the operating point at C for the pumps operating in parallel: Q_{CD} at H_C (point 3). Note that, in this case, point 3 doesn't coincide with point 1. But the resulting capacity will be tolerably higher than 4,000 GPM.

12. Draw a horizontal line from point 3 (head H_C) to intersect curves $(H-C)-H_{f\ AC}$, and $(H-C)-H_{f\ BC}$, which, as indicated above, represent the performance curves of each pump taken at the junction point C. Obtain points 4 and 5.

13. Read off the individual capacities: Pump B$=Q_{BC}$; Pump A$=Q_{AC}$. By construction, $Q_{BC}+Q_{AC}=Q_{CD}$.

14. Draw vertical lines from points 4 and 5 to intersect the head-capacity curves of each pump, and obtain points 6 and 7, which represent the true pump operating points. These operating points are different because the pumps are not equidistant from junction point C.

15. Pump operating points (for parallel operation):

Pump A: point 7—Q_{AC} at H_{A2}

Pump B: point 6—Q_{BC} at H_{B2}

16. For individual pump operations, locate intersections of $H_{f\ BCD}$ and $H_{f\ ACD}$ and the pump curve (H−C), and obtain points 9 and 10; 9′ and 10′ are heads at C for single pump operations.

17. Operating points for individual operation: Pump A—point 10—Q_{A1} at H_{A1}; Pump B—point 9—Q_{B1} at H_{B1}.

18. Draw vertical line at Q_{CD} to intersect the combined pump curve at point 8, representing the common total operating point (Q_{CD} at H_M'). H_M' is the average head for the two pumps: $H_M'=(H_{B2}+H_{A2})/2$. The combined curve for junction $C=2(H-C)-(H_{fAC}+H_{fBC})/2$, as indicated in item 10. Note that points 1 and 2 shifted to 3 and 8.

Solution:

1. Compute and plot the system head curves (assume $f_{AC}=f_{BC}=f_{CD}=0.013$):

$$H_{fAC}=\frac{0.013\times 600}{10^5\times 32.18}\times Q^2_{AC}=Q^2_{AC}/412{,}000$$

$$H_{fBC}=\frac{0.013\times 1{,}400}{10^5\times 32.18}\times Q^2_{BC}=Q^2_{BC}/177{,}000$$

$$H_{fCD}=\frac{0.013\times 12{,}000}{12^5\times 32.18}\times Q^2_{CD}=Q^2_{CD}/51{,}300$$

$H_{fACD}=H_{fAC}+H_{fCD}=Q^2_A/45{,}600$ (single pump)

$H_{fBCD}=H_{fBC}+H_{fCD}=Q^2_B/39{,}800$ (single pump).

Points for the System-Head Curves:

Q (GPM)	1,000	2,000	3,000	4,000
H_{fAC} (FT)	2.5	9.7	22	39
H_{fBC} (FT)	5.7	23	51	90
H_{fCD} (FT)	20	78	175	312
H_{fACD} (FT)	22	88	197	350
H_{fBCD} (FT)	25	101	226	402

Refer to Figures 16.06 and 16.07.

Results as read off fig. 16.07:

Point 8 = Operating point for both pumps in parallel: 4,070 GPM at an average head of 340 FT. Referring to points 6 and 7, 347+333=680; 680/2=340 FT. The point deviates slightly from the assumed point 2.

Point 8′ = 2,035 GPM at 340 FT, the hypothetical operating point for the two pumps, operating in parallel, and installed at a distance of (600+1,400)/2=1,000 FT away from C.

Point 6 = True operating point for pump B: 1,880 GPM at 347 FT., when both pumps are operating simultaneously.

Point 7 = True operating point for pump A: 2,190 GPM at 333 FT., when both pumps are operating simultaneously.

Point 3 = Head at junction C for simultaneous operation: 323 FT. Branch loss for pump A=333−323=10 FT, and branch loss for pump B=347−323=24 FT. The average branch loss (10+24)/2=17 FT. The estimated average branch loss was 30 FT.
17+323 (per point 3)=340 FT per point 8.

Point 9 = Pump B operating singly: 3,235 GPM at 263 FT (80% of 4,000 GPM).

Point 9′ = Head at junction C for pump B operating singly: 200 FT. Branch loss in this case: 263−200=63 FT=head at point 9 minus head at point 9′.

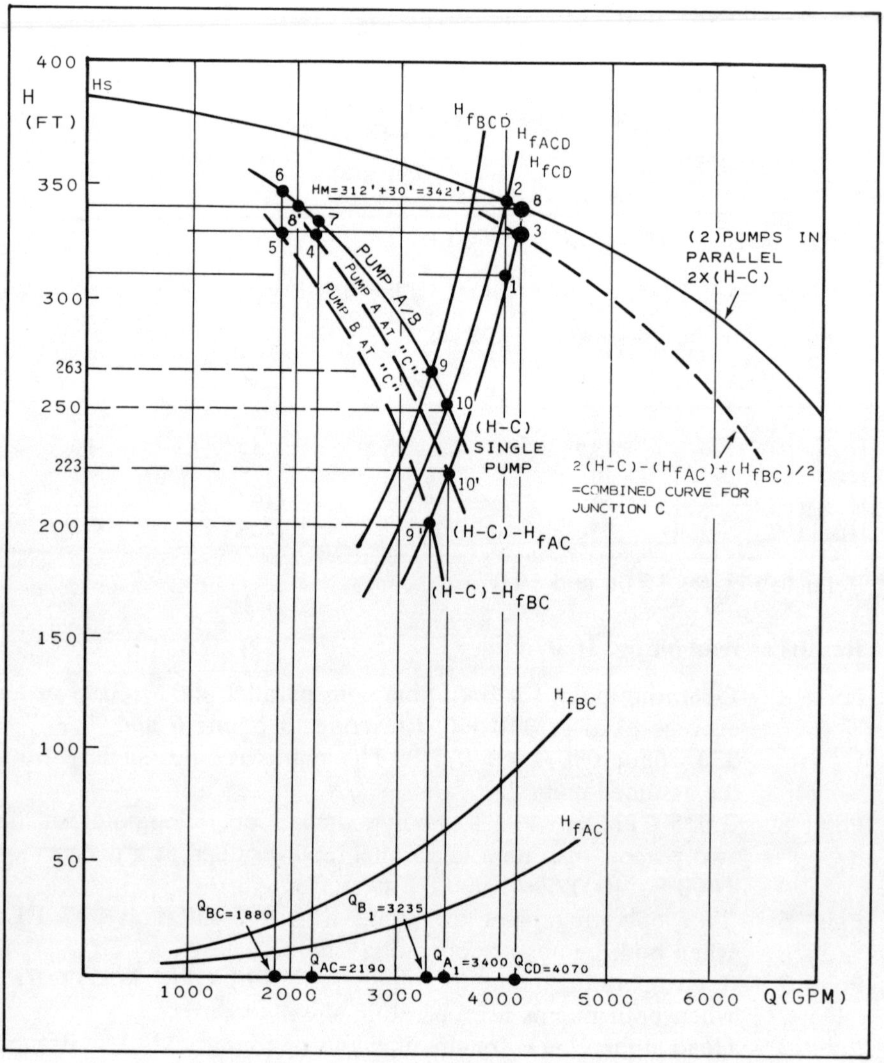

FIG. 16.07

Point 10 = Pump A operating singly: 3,400 GPM at 250 FT. (85% of 4,000 GPM). Branch loss in this case: 250 − 223 = 27 FT. = head at point 10 minus head at point 10′.

Comments: The pumps have to be capable of providing stable operation at points 9 and 10. The above values will deviate slightly from computed ones which is to be expected when values are read off graphs.

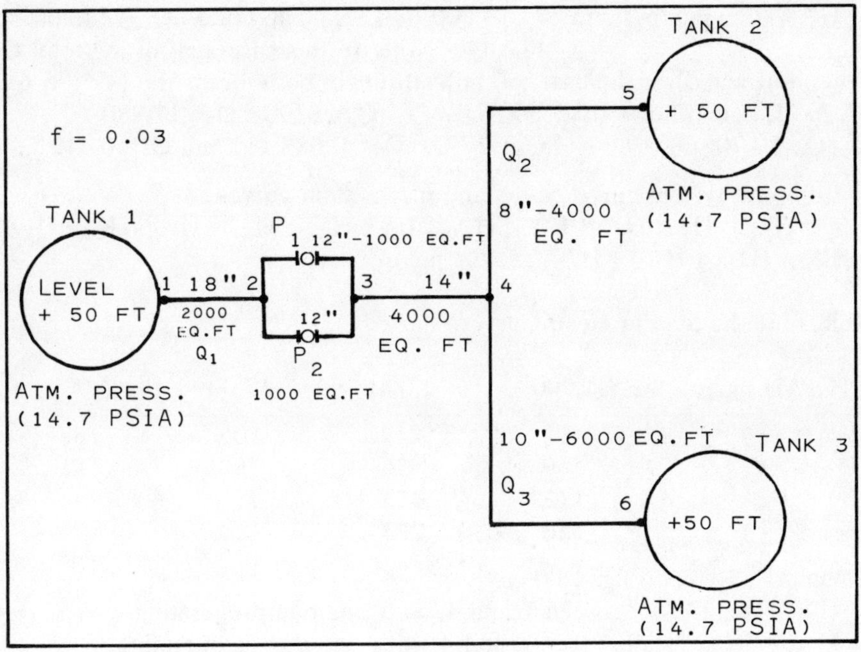

FIG. 16.08

■ EXAMPLE 16.4

Select two identical pumps to operate in an arrangement per fig. 16.08 as follows:

A. In parallel, to transfer 4,200 GPM of oil to tanks 2 and 3 simultaneously.

B. Singly, to transfer oil to both tanks simultaneously, or transfer oil to tank 2 or to tank 3.

Determine all capacities and heads.

Solution:

System head curves are as follows:

$H_{1\text{-}2} = 0.03 \times 2{,}000 \times Q^2/(18^5 \times 32.18) = Q^2/1{,}013{,}000$

$H_{2\text{-}3} = 0.03 \times 1{,}000 \times Q^2/(12^5 \times 32.18) = Q^2/267{,}000$

$H_{3\text{-}4} = 0.03 \times 4{,}000 \times Q^2/(14^5 \times 32.18) = Q^2/144{,}200$

$H_{4\text{-}5} = 0.03 \times 4{,}000 \times Q^2/(8^5 \times 32.18) = Q^2/8{,}790$

$H_{4\text{-}6} = 0.03 \times 6{,}000 \times Q^2/(10^5 \times 32.18) = Q^2/17{,}880$

$H'_{1\text{-}4}$ (one pump) $= Q^2[1/1{,}013{,}000 + 1/267{,}000 + 1/144{,}200]$

$= Q^2/85{,}710$

$H''_{1\text{-}4}$ (two pumps) $=Q^2/1{,}013{,}000+(0.5Q)^2/267{,}000+Q^2/144{,}200$ $=Q^2/112{,}890$. Note: In this case, half of the total Q flows through 2-3. In case of flow through both branches (4-5; 5-6), $H_{4\text{-}5}=H_{4\text{-}6}$; it follows that: $Q_1=Q_2+Q_3$; $Q_2^2/8790=Q_3^2/17{,}880$; $Q_2/Q_3=0.70$; $Q_1=0.7Q_3+Q_3=1.7Q_3$; $Q_3=0.588\ Q_1$; and $Q_2=0.412\ Q_1$.

Compute system curve points and plot system curves $H'_{1\text{-}4}$; $H''_{1\text{-}4}$; $H_{4\text{-}5}$; $H_{4\text{-}5}+H'_{1\text{-}4}$; $H_{4\text{-}6}$; $H_{4\text{-}6}+H'_{1\text{-}4}$; $H_{4\text{-}5}+H_{4\text{-}6}$; $(H_{4\text{-}5}+H_{4\text{-}6})+H'_{1\text{-}4}$; $(H_{4\text{-}5}+H_{4\text{-}6})+H''_{1\text{-}4}$. Select pump curves.

Points for the System Head Curves:

Q(GPM)	1,000	2,000	3,000	4,000
$H'_{1\text{-}4}$ (FT)	12	47	105	187
$H''_{1\text{-}4}$ (FT)	9	35	80	141
$H_{4\text{-}5}$ (FT)	113	455	. . .	. . .
$H_{4\text{-}6}$ (FT)	56	223	. . .	. . .

Summary (Refer to fig. 16.09):

$H'_{1\text{-}4}$=head loss between 1 and 4, with one pump operating.

$H''_{1\text{-}4}$=head loss between 1 and 4, with two pumps operating.

$H_{4\text{-}5}$=head loss in branch 4-5.

$H_{4\text{-}6}$=head loss in branch 4-6.

$H'_{1\text{-}4}+H_{4\text{-}5}$=head loss between 1-5, with one pump operating (heads added for same capacities).

$H'_{1\text{-}4}+H_{4\text{-}6}$=head loss between 1-6, with one pump operating (heads added for same capacities).

$H_{4\text{-}5}+H_{4\text{-}6}$=combined head loss in branches. (capacities added for same heads).

$(H_{4\text{-}5}+H_{4\text{-}6})+H'_{1\text{-}4}$=Total head loss for one pump to both branches. (heads added for same capacities).

$(H_{4\text{-}5}+H_{4\text{-}6})+H''_{1\text{-}4}$=Total head loss for two pumps to both branches. (heads added for same capacities).

Summary of results, read off fig. 16.09:

Point 1=Required point of operation for two pumps: 4,200 GPM at 499 FT., 2,475 GPM in branch 4-6, and 1,725 GPM in branch 4-5. Operating point for each pump: 2,100 GPM at 499 FT. Points 2 and 3 indicate distribution. Head at junction 4=339 FT (point 1′).

Point 4=One pump singly to branch 4-5. Operating point: 2,000 GPM at 502 FT.

Point 5=One pump singly to branch 4-6. Operating point: 2,580 GPM at 450 FT.

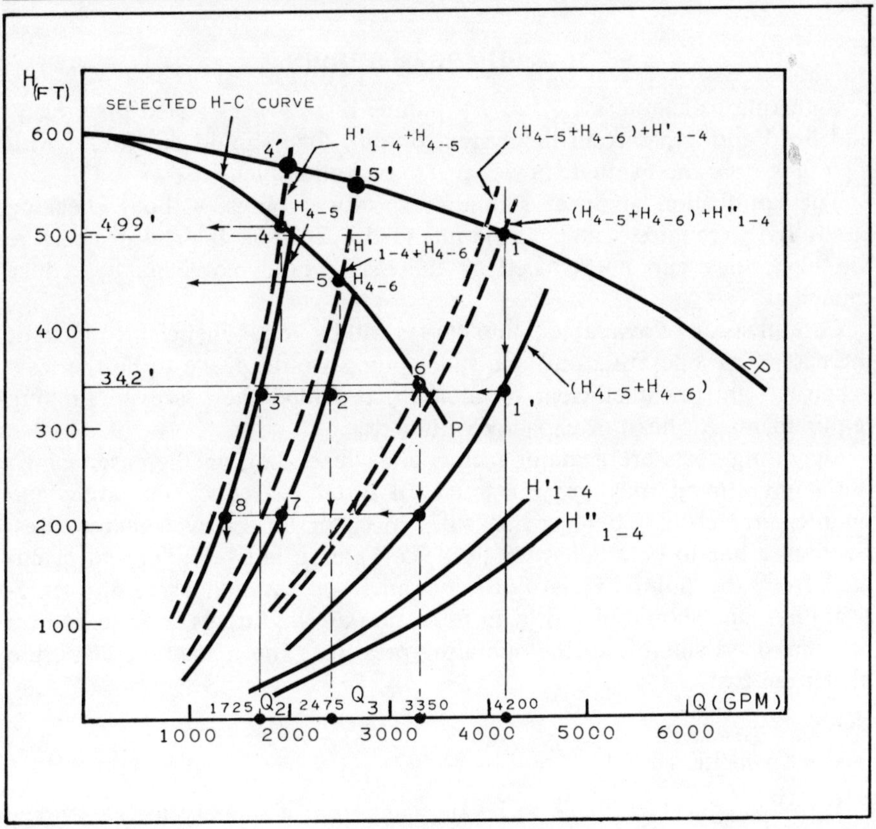

FIG. 16.09

Point 6 = One pump to both branches. Operating point: 3,350 GPM at 342 FT. Distribution in branches per points 7 and 8: 1,965 GPM and 1,385 GPM. Head at Junction 4 = 210 FT.

Note that operating both pumps in parallel to one or the other tank only doesn't for practical purposes, add any capacity to the one of a single pump. (points 4′ and 5′).

The relations between capacities are as follows:

$Q_1 = Q_2 + Q_3 = 4{,}200 = 1{,}725 + 2{,}475$

$Q_3/Q_1 = 0.588 = 2{,}475/4{,}200$, as computed before.

$Q_2/Q_1 = 0.412 = 1{,}725/4{,}200$, as computed before.

Note: The selected pumps have to provide stable operation at point 6 (3,350 GPM).

16.2 PUMPS IN SERIES

Referring to chapters 14 and 15, pumps in series, as stated previously, add head and capacity to the system output. One should be careful with high pressures and evaluate power cost versus equipment cost.

The application engineer shouldn't specify a pump without checking carefully the requirements and the intended operation. If pressures will be too high, they can be reduced by increasing pipe diameters or reduce capacities.

Careful study of available pump curves will be very beneficial. In most of the cases, one operating and one spare pump would prove to be the least expensive and most efficient installation, provided there are no limiting requirements for head or capacity variations.

Operating costs are a major factor, and, depending on the case, a high initial investment may be warranted, in order to reduce operating and maintenance costs. There are no rules to cover all, or even many cases. Each case has to be studied for itself, considered individually, and evaluated from the point of view of economics, operational ease, and practicability. One should also bear in mind that all the fittings have to be of a pressure class suitable to the operating pressure. The higher the class, the higher the cost.

■ EXAMPLE 16.5

Petroleum products have to be supplied from a refinery to the storage facilities of three customers in three different locations. Transfers will be individual, to one customer at a time. Required flowrate range is 3,000-4,000 GPM. A 12″ pipeline connects the refinery with the customers' storage facilities, as indicated on fig. 16.10.

Compute the individual heads and horsepower required upon selection of a suitable curve for the three pumps to operate in series. Static heads can be neglected. Average product viscosity: 5 centistokes; specific gravity: 1.1. Discuss the solutions and alternatives.

Solution:

Determine f:

$RE = 3{,}162 \times GPM/(D'' \times Cs) = 3{,}162 \times 4{,}000/(12 \times 5) = 210{,}800$

$f = 0.017$ (new pipes; from charts)

The system head curves:

$H_{R\text{-}1} = 0.017 \times 18{,}000 \times Q^2/(12^5 \times 32.18) = Q^2/26{,}170$

$H_{R\text{-}2} = 0.017 \times 33{,}000 \times Q^2/(12^5 \times 32.18) = Q^2/14{,}275$

$H_{R\text{-}3} = 0.017 \times 88{,}000 \times Q^2/(12^5 \times 32.18) = Q^2/5{,}353$

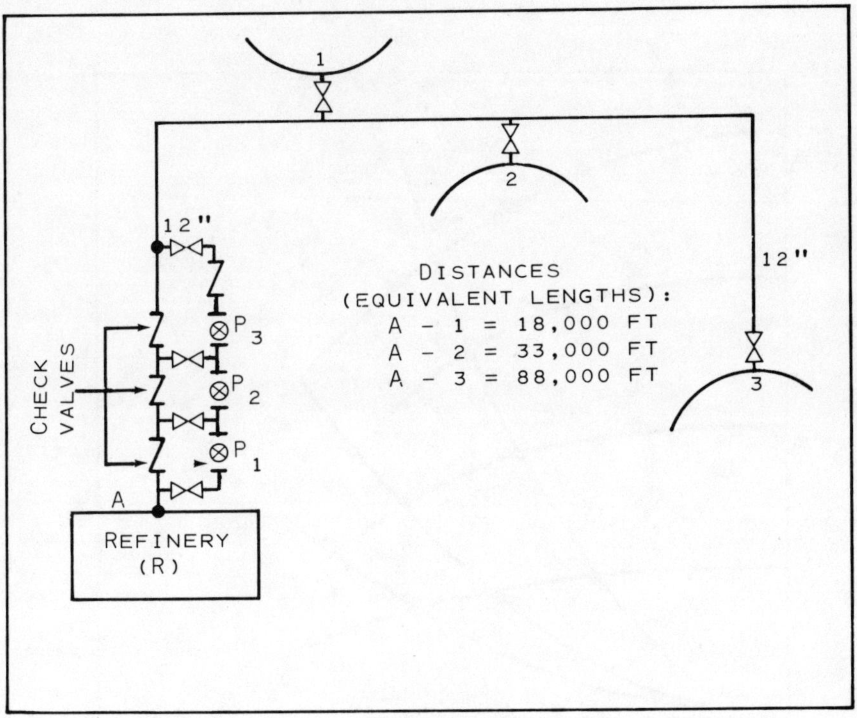

FIG. 16.10

Points for the System Head Curves:

Q (GPM)	1,000	2,000	3,000	4,000	5,000
$H_{R\text{-}1}$ (FT)	38	152	344	610	955
$H_{R\text{-}2}$ (FT)	70	280	630	1,120	1,750
$H_{R\text{-}3}$ (FT)	186	747	1,680	2,990	4,670

Refer to Fig. 16.11.

$$BHP = \frac{H_{(FT)} \times GPM \times sp.\ gr.}{3{,}960 \times \text{pump efficiency (\%–fraction)}}$$

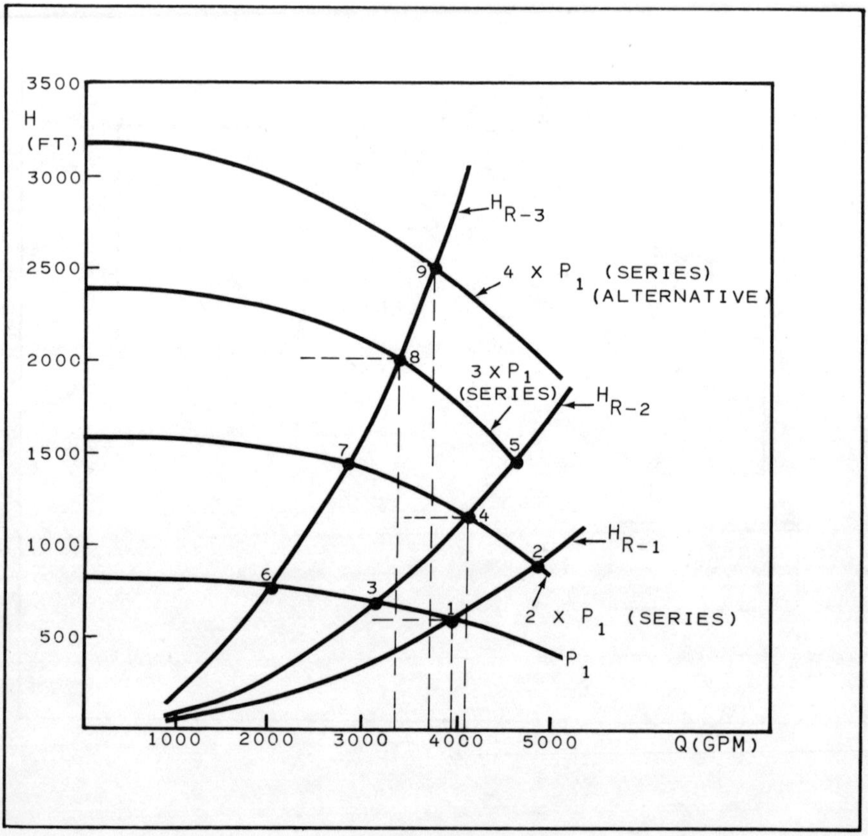
3500
H
(FT)
3000
2500
2000
1500
1000
500
1000
2000
3000
4000
5000
Q(GPM)
H_{R-3}
4 X P_1 (SERIES) (ALTERNATIVE)
3 X P_1 (SERIES)
H_{R-2}
H_{R-1}
2 X P_1 (SERIES)
P_1
1
2
3
4
5
6
7
8
9

FIG. 16.11

Results as read off fig. 16.11:

	TOTAL				PER PUMP					
POINT	GPM	FT	PSIG	BHP	GPM	FT	PSIG	BHP	%EFF.	NO. OF PUMPS
				TO CUSTOMER 1:						
1	3,950	600	285	823	3,950	600	285	823	80	1
2	4,820	900	428	1,772	4,820	450	214	886	68	2
				TO CUSTOMER 2:						
3	3,120	700	333	800	3,120	700	333	800	76	1
4	4,100	1,180	562	1,618	4,100	590	281	809	83	2
5	4,600	1,450	690	2,640	4,600	483	230	880	70	3
				TO CUSTOMER 3:						
6	2,000	800	381	673	2,000	800	381	673	66	1
7	2,820	1,440	685	1,636	2,820	720	343	818	69	2
8	3,350	2,020	962	2,412	3,350	673	320	803	78	3
*9	3,700	2,520	1,200	3,280	3,700	630	300	820	79	4

*Alternative.

Comments: The pressures in the 12″ line are high, as indicated. This means that high-strength pipe and fittings will have to be selected. One could obtain lower pressures by increasing the pipe diameter to 14″ or 16″. An economic analysis is required to determine the optimum diameter.

In case the capacity range indicated (3,000-4,000 GPM) for each customer has to be maintained, and if pumping to the three customers successively represents a 24-hour a day operation, one would have to install a fourth spare pump, since with two pumps (one pump down) customer 3 cannot be satisfied.

Under normal conditions, one should be able to satisfy the three customers with three pumps, even if one pump is down for maintenance or repairs.

The alternative with 4 pumps in series is included in the example to bring up the ensuing inefficiency: per point 8, three pumps to customer 3 would supply 3,350 GPM which is within the required range, at the expense of 3×803=2,412 BHP, or 0.72 BHP per GPM. With 4 pumps (point 9), the total capacity would amount to 3,700 GPM, an increase of 10.4%, whereas the increase of BHP amounts to 3,280/2,412=1.36, or 36%. In other words, for an increase of 350 GPM, one would expend additional 868 HP.

For an installation of this type, all pumps should be identical, in order to store identical spare parts, and to have identical and less expensive maintenance.

The application engineer should shop for efficient and flexible pump curves to satisfy the conditions in an optimum way. Upon the selection of the curves, the friction factors (f) have to be recomputed which will allow a precise recomputation of the BHP required for the selection of the driver.

■ EXAMPLE 16:6

Per fig. 16.12, chemicals have to be transferred either to a close-by facility, or, occasionally, to a distant one. Two pumps will operate in series, or singly, as required. Select suitable pump curves and comment. Viscosity of products = 10 cs, and specific gravity = 0.80. Required capacity to facility 1 or 2: 6,000 GPM. Friction factor = 0.020.

The system-head curves:

$$H_{AB1} = \text{Static Head} + f\frac{L}{D^5} \times \frac{Q^2}{32.18} = 100 + 0.02\frac{5000}{16^5} \times \frac{Q^2}{32.18}$$

$$= 100 + Q^2/337{,}400$$

$$H_{AB2} = 60 + 0.02\frac{28{,}000}{16^5} \times \frac{Q^2}{32.18} = 60 + Q^2/60{,}250$$

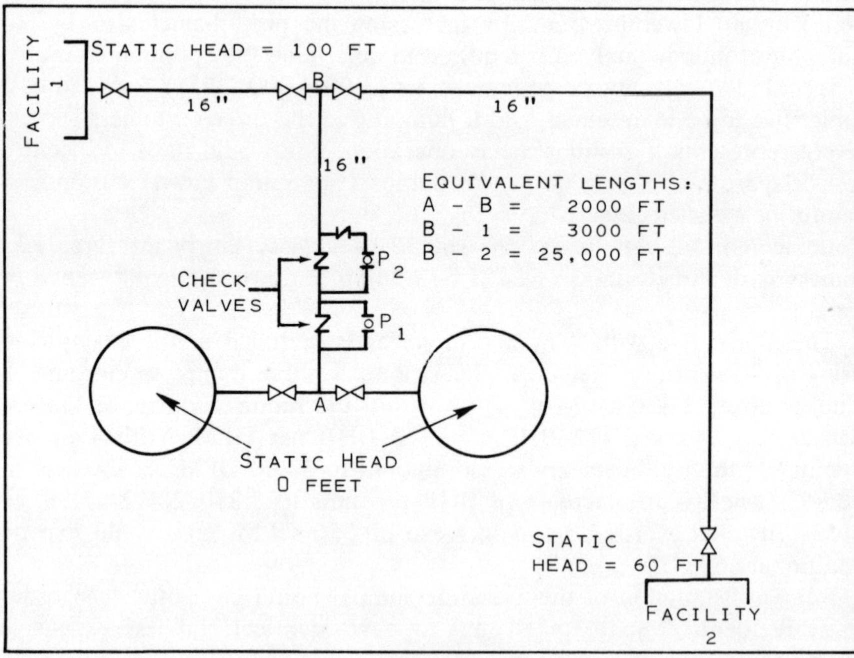

FIG. 16.12

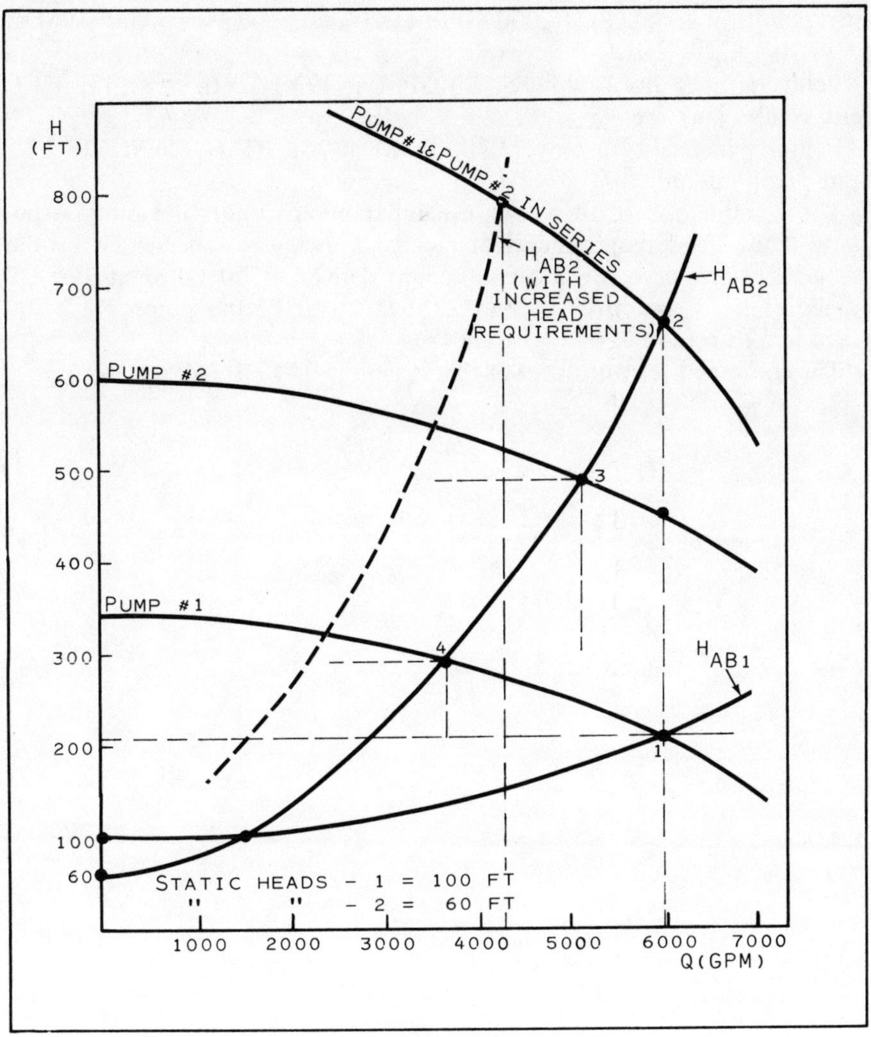

FIG. 16.13

Compute points and plot system head curves (Fig. 16.13).

Results and comments:

Point 1: to facility 1 only: 6,000 GPM at 207 FT (fr. loss: 107 FT); (one pump operating—P_1).

Point 2: to facility 2 only: 6,000 GPM at 657 FT (fr. loss: 597 FT);

(two pumps in series; P_1=6,000 GPM at 207 FT, and P_2=6,000 GPM at 450 FT).

Point 3: to facility 2 only: 5,100 GPM at 493 FT (fr. loss: 433 FT); (one pump operating—P_2).

Point 4: to facility 2 only: 3,700 GPM at 288 FT (fr. loss: 228 FT) (one pump operating—P_1).

The combination of these two dissimilar pumps offers flexibility, especially if the head requirements at facility 2 increase, represented by the dashed system curve. Both pumps could deliver 4,300 GPM at 790 FT combined head (pump P_1: 4,300 GPM at 268 FT, and pump P_2: 4,300 GPM at 522 FT).

The combined shut-off head of this system is 940 FT (600′+340′).

17.

Arrangement of Pumps for Operation in Parallel or Series as Function of the System Head Curve

When using multiple pumps, careful consideration must be given to each particular case before deciding whether to arrange the pumps in parallel or in series. The plotting of the system-head curve for the specific application with all its variations during the operation is mandatory. By superimposing suitable pump curves on the system head curves, one can judge whether one should decide for operation in parallel or in series.

■ EXAMPLE 17:1

Discuss in general terms the operation of multiple pumps for a fixed, flat system head curve. Refer to fig. 17.01.

Two pumps in parallel offer these advantages:

- Capacity has increased by 50% (point 2).
- Head has increased, but only slightly.
- Pump operating points are in the vicinity of higher efficiencies.

Two pumps in series would be operating inefficiently:

- Shut-off head would double. This means high heads during start-up.
- Capacity increases by only 35% (point 3).
- Pump operating points are in the vicinity of lesser efficiencies (3 and 3′).

In conclusion, pumps operating in parallel for a flat system will be more efficient.

■ EXAMPLE 17:2

Discuss in general terms the operation of multiple pumps for a variable flat system.

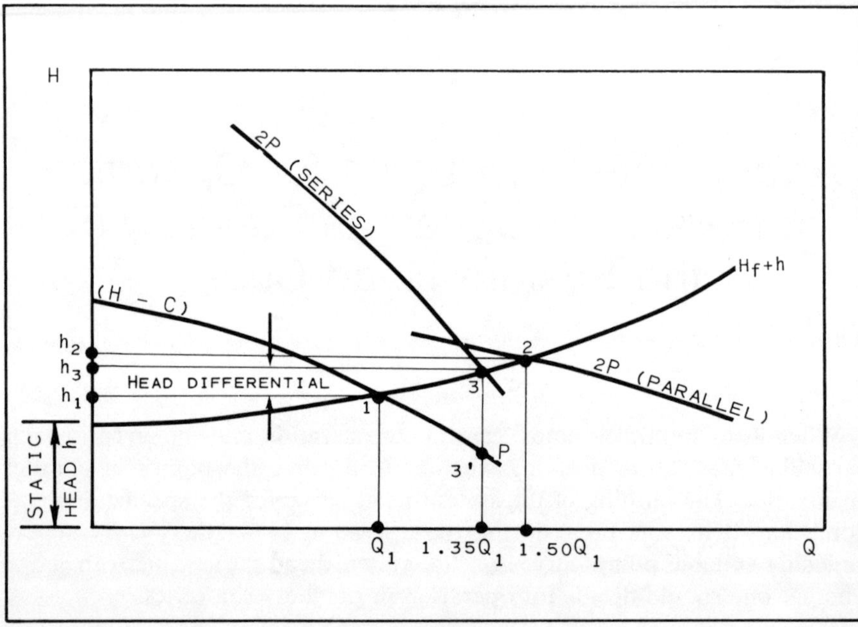

FIG. 17.01

In this case, individual transfers have to be made to different consumers. Refer to fig. 17.02.

Case 1. Consumers 1 and 2 require frequent transfers; consumer 3 requires occasional transfers.

Alternative 1.1: Three pumps in parallel. Two pumps in parallel for consumers 1 or 2 (points 1 and point 2 on fig. 17.02), three pumps in parallel for consumer 3, in case Q_{3a} is satisfactory (point 3a). This solution is feasible for small, and therefore, inexpensive pumps, with higher heads.

Alternative 1.2: Two pumps arranged to operate either in parallel or in series, per schematic arrangement on fig. 17.03.

This arrangement will be more economical if larger capacities, and not too high heads are required. Cost is a factor when comparing both alternatives.

Operating points are: Points 1 and 2 for consumers 1 and 2, as above (two pumps in parallel), and point 3, operation in series, for consumer 3. An optimization study covering cost, power consumption, etc. will be required.

Case 2. All three consumers require frequent transfers on an individual basis. Two pumps in series may prove attractive, provided capacities are satisfactory (Operating points are: 1′, 2′, 3).

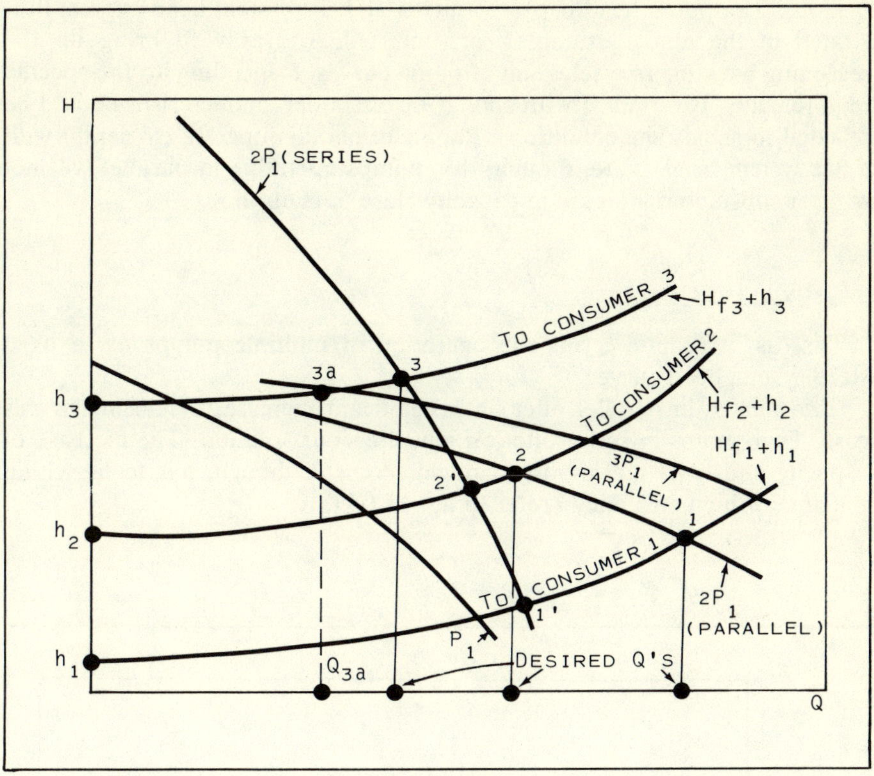

FIG. 17.02

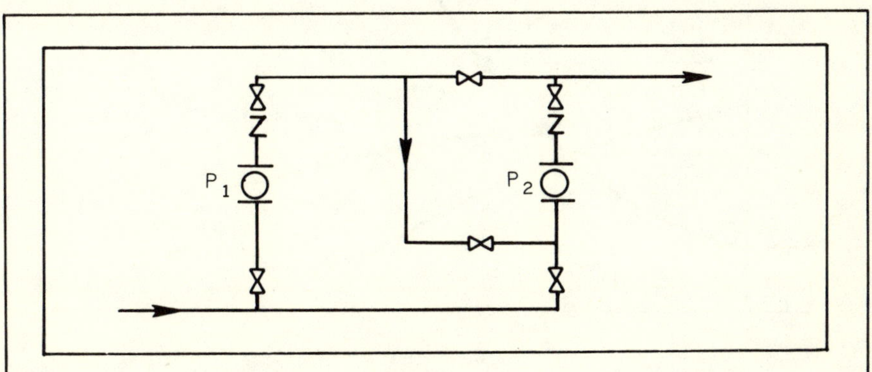

FIG. 17.03

Note: The pump head capacity curves and the system head curves, illustrated in the above examples, are selected arbitrarily to bring up the reasoning backing up a selection of pump curves. Depending on the specific requirements, two pumps with similar or dissimilar characteristics could be selected to satisfy the conditions. Pumps in parallel operate efficiently with a flat system head. Note, though, that pumps operating in parallel will not register substantial increase in capacity close to shut-off.

■ EXAMPLE 17:3

Discuss in general terms the operation of multiple pumps for a fixed steep system head curve.

Two pumps in parallel offer an insignificant increase in capacity in this case. Two pumps in series offer an advantageous solution. The increase of capacity and head are almost identical. Point 2, though, has to be within the area of high efficiency (refer to fig. 17.04).

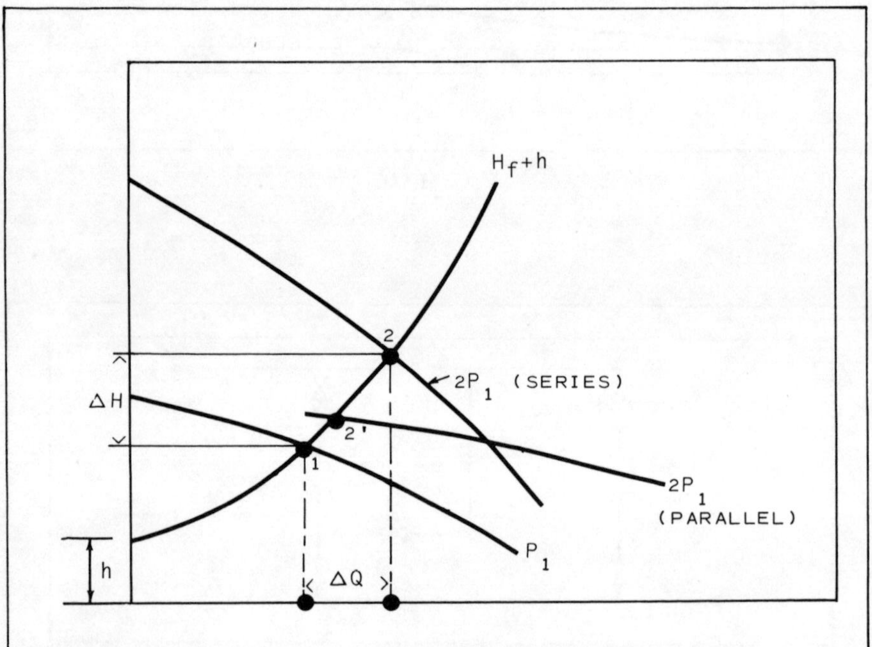

FIG. 17.04

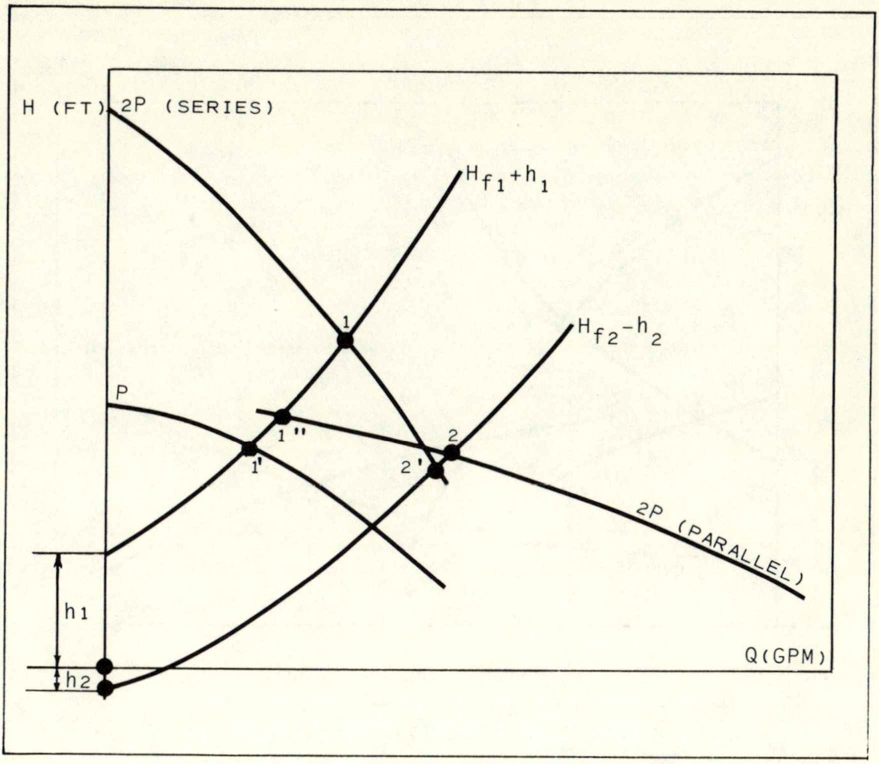

FIG. 17.05

▪ EXAMPLE 17:4

Discuss in general terms the operation of multiple pumps for a variable steep system head curve. Refer to fig. 17.05.

Parallel operation would not satisfy requirements per point 1, and would be a very inefficient combination for point 1″, as compared to the output of a single pump per point 1′. In series, one can achieve operating points 1 and 2′. An arrangement in series is more advantageous since it almost satisfies both points which would be satisfied by a parallel arrangement (1″ and 2), and, in addition to those, also point 1.

▪ EXAMPLE 17:5

Discuss in general terms the operation of multiple pumps for two fixed system heads: one steep and one flat (individual transfers) (Refer to fig. 17.06).

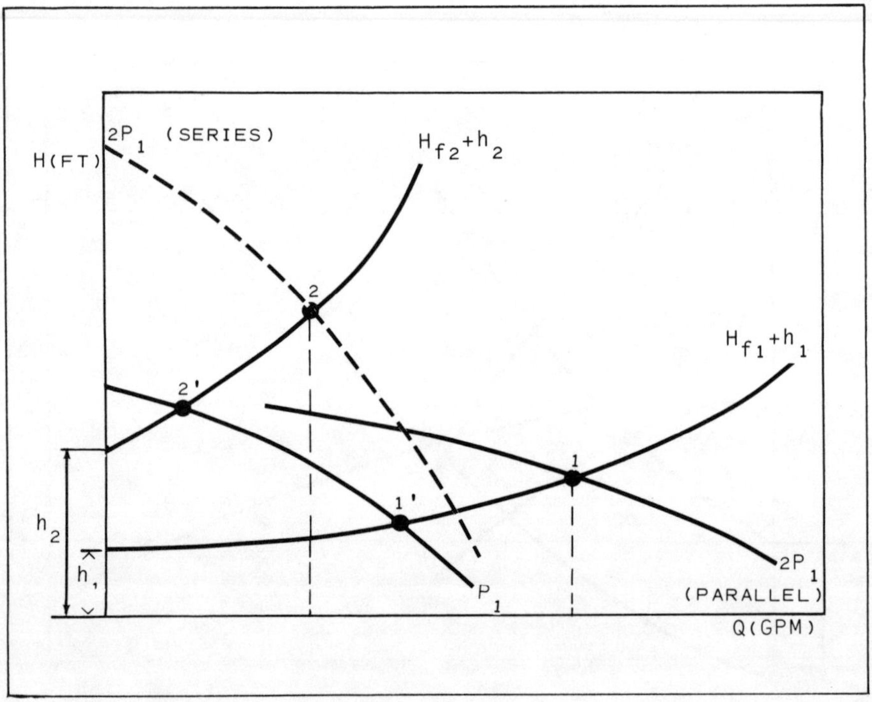

FIG. 17.06

Case 1. Identical pumps. The advantages are identical spare parts, a symmetrical arrangement, and identical space requirements. Point 1 requires operation in parallel, and Point 2 in series. Depending on the frequency and importance of the service, one could have a parallel-series arrangement (see fig. 17.03). For uninterrupted operation, in case capacities have to be maintained, one would need a third identical pump as a spare, installed in the same manner.

Case 2. Dissimiliar pumps in parallel. This arrangement will prove advantageous in case occasional decreases in capacity can be tolerated when one pump will be out of service (refer to fig. 17.07).

P_1 by itself will produce 2b and 1a, while P_2 by itself will produce 2 and 1b.

The combined curve for both pumps operating in parallel (P_2-C-D) will produce 2a and 1, the latter more than doubling the capacity. For this arrangement, point 1 should be considered the most frequent requirement in order to justify two pumps. Otherwise one could get by with a single pump and a spare.

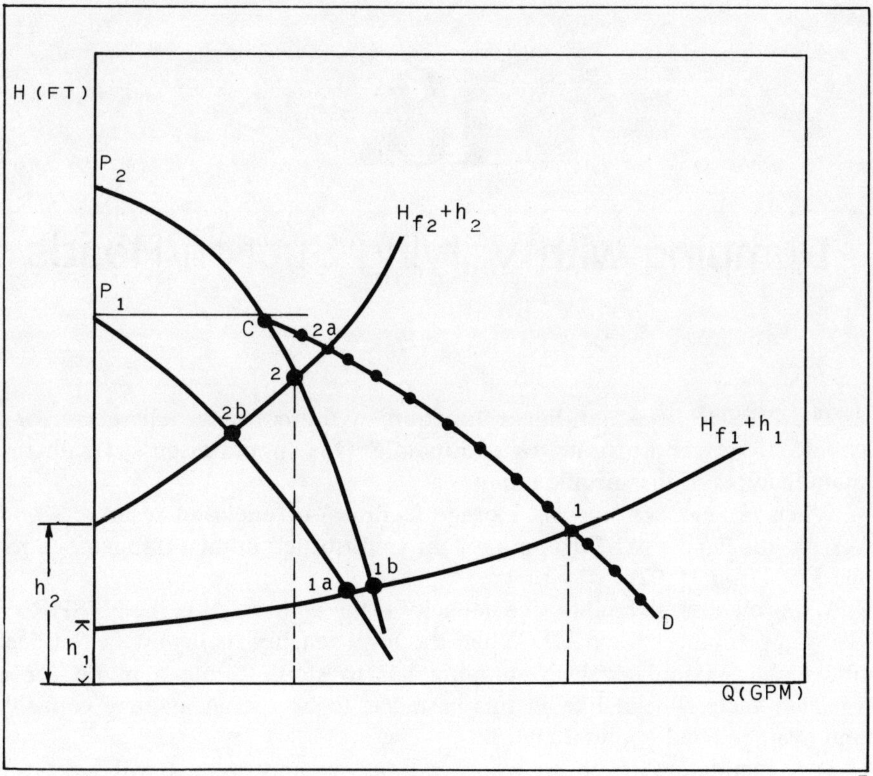

FIG. 17.07

18.

Pumping with Varying Suction Heads

The NPSHR is a non-linear function of the capacity. Therefore, one should not try and prorate, or extrapolate. One should discuss it with the manufacturer of the specific pump.

When taking suction from storage facilities, or enclosed reservoirs and vessels, the head will be dropping. This is illustrated in the arrangement per fig. 18.01 and 18.02.

With full storage facility, the capacity is the highest, so is the NPSHR of the pump (point 1 for h_{max}). When the level reaches its lowest (a few feet above the suction nozzle), pumping has to stop. Rating a pump for a variable suction head like in this case has to be for an average capacity and average head (point 1a).

It is mandatory to check whether large-capacity pumps will have the

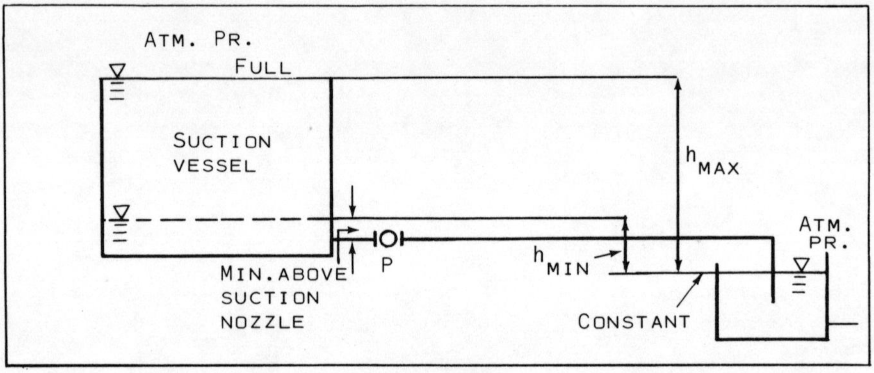

FIG. 18.01

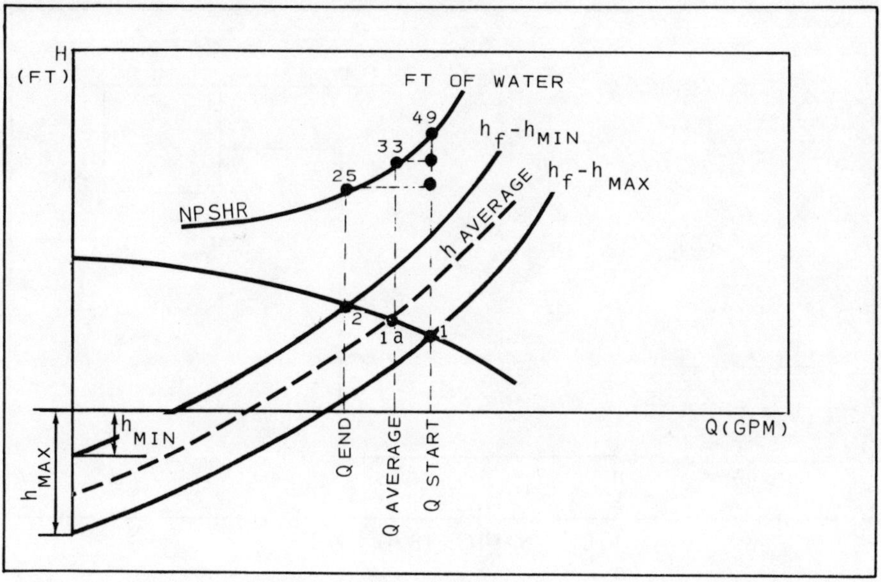

FIG. 18.02

NPSHR satisfied for the maximum capacity. They may require between 40 and 60 feet of water toward the end of their curve.

If for the particular case per fig. 18.02 $h_{max}=25$ FT, and the friction losses in the suction line amount to 12 FT, then, neglecting vapor pressure, NPSHA$=25-12+34=47$ FT (for atmospheric pressure$=14.7$ PSIA and Sp. gr.$=1.0$). It is not sufficient, since the NPSHR is larger than 47 FT (49 FT). The conditions for point 1a are satisfactory for $h_{min}=5$ FT, and $h_{av}=15$ FT. For a friction loss of 8 FT, the NPSHA$=41$ FT, which is larger than the 33 FT required per fig. 18.02.

The pump would cavitate for a while, upon starting with full tank, a condition to be avoided by means of proper investigation and pump-curve selection. Very important, too, is to check the BHP requirements at point 1 and specify the electric motor accordingly.

For cases for which the NPSHR is critical, one can use vertical, canned pumps. In this arrangement, the suction is below the bottom of the tank. A canned pump also requires a minimum submergence in order to avoid vortexes. If the tank, or vessel can be elevated, the pump can be installed under it in order to generate flooded suction conditions.

Varying the suction head for systems with high discharge head can cause problems if the arrangement and the selection of the pump curve were not paid attention to as one should. Consider the arrangement per fig. 18.03 and the characteristics on fig. 18.04.

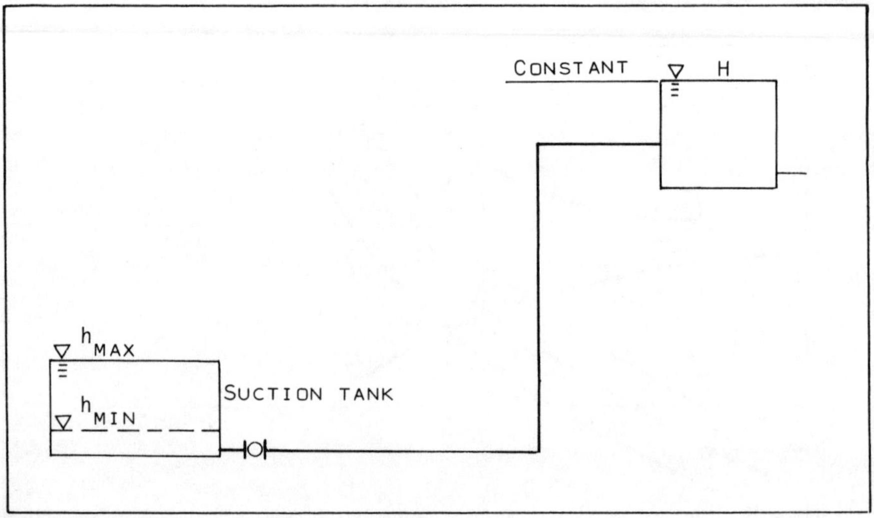

FIG. 18.03

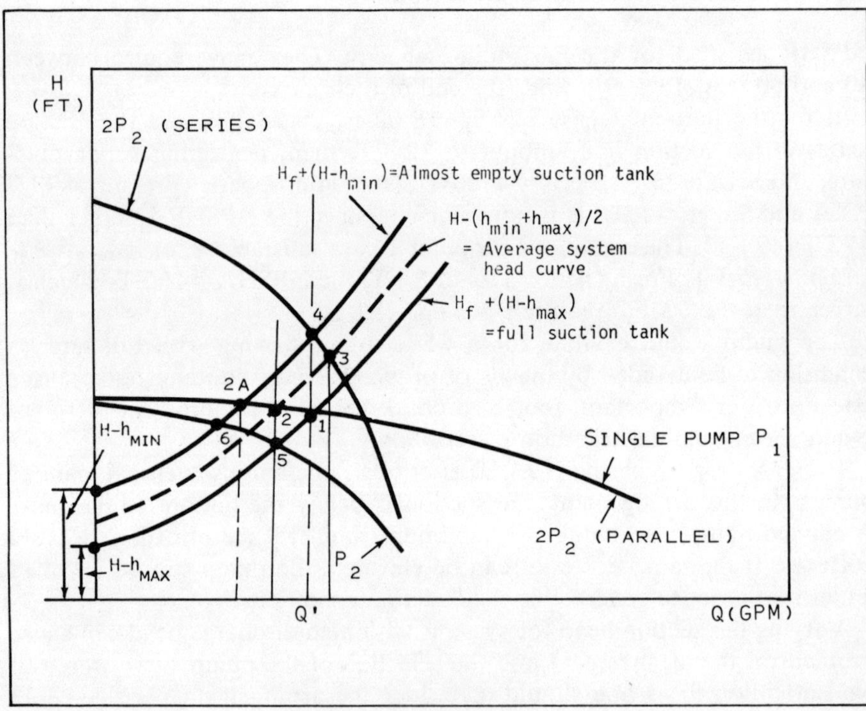

FIG. 18.04

Case 1. One single pump P_1 with operating points 1, 2, and 2a, with liquid level in the suction tank varying from full to almost empty. Capacity for 2a is approximately 70% of the capacity for point 1. The curve is flat and affects the change in capacity substantially.

Case 2. Two identical pumps P_2 are arranged to operate alternately in series or parallel. They operate in parallel from points 1 to 2, then switch to operation in series, points 3 to 4.

During the total time of operation, the average capacity will be in the vicinity of Q′ which represents the design capacity. With one pump down, the operating points will be 5 and 6.

If the process doesn't permit lower than design capacities on the average, then a third pump has to be installed to serve as spare. Note that a series/parallel arrangement requires more valving and piping (initial cost higher), as well as more operating time. Therefore, it will be justifiable only for cases where large capacities are involved.

19.

The Economics of Pump Curve Variation

The reader is advised to read again chapter 11. Upon learning in detail the process requirements for a pumping system and determination of the system-head curve or curves, the application engineer has to select the number of pumps, the type of combined operation, if any, and the controls to meet the specific requirements.

In previous chapters, we discussed in detail operations in parallel and in series with identical and dissimilar pumps, and brought up the advantages and disadvantages of each system. The pump head-capacity curve was considered constant (constant impeller diameter and pump speed).

There are cases where variable pump head-capacity curves have to be considered. The best way to vary a pump curve is by varying the pump speed. Note that the pump curve will shift if we replace the impeller with one of a different diameter. But this is only done for cases of more permanent nature.

Variation of pump speed will provide operation flexibility, efficiency, and energy savings. One should bear in mind, though, that systems permitting speed variation cost more and require additional maintenance, all of which should be taken into consideration when preparing an economic analysis.

Drivers which will permit speed variation are: steam turbines, gas turbines, gasoline engines, diesel engines, and variable-speed electric motors. With constant-speed motors one can use hydraulic couplings, belt drives, and magnetic drives to vary the speed.

Generally speaking, it would be of great advantage if one could vary the pump speed at will. This is not done very often because of the increase of the initial and the maintenance costs.

Steam turbines are used often if high pressure steam is available anyway, as for example in a power plant or refinery. Tapping the main steam source would be economical, especially when one has to operate large pumps. Step-down reduction gears are usually required. Gas turbines and internal combustion engines are used in locations where electric power is unavailable.

Variable speed motors are preferred, because, like any other electric motor, they require the least maintenance and occupy the least space. They are also the most efficient. Where variable speed is mandatory, one can use hydraulic couplings or magnetic drives, coupled to constant-speed motors. Cost and space, as well maintenance requirements, though, are high.

It can be concluded that, from the point of view of cost, space, and maintenance requirements, the electric-motor-driven pump represents the most efficient combination. For critical services (fire protection, flood control, etc.), the spare pumps have non-electric drivers.

When deciding on variable speed, the pump manufacturer must be consulted for the range of allowable speed variation. The speed at which the pump will operate most of the time should be in the vicinity of the maximum speed for the specific pump.

High-pressure boiler-feed pumps (heads of up to 10,000 ft) are multistage by construction and require high speeds, up to 10,000 RPM. These are achieved by using turbines with step-up gears. In this range and size, variation of speed is of major importance.

■ EXAMPLE 19:1

Discuss variable speed as applied to the system per fig. 19.01.

For this system:

$$N_2/N_1 = (H_2/H_1)^{1/2};\; N_2 = N_1\left(\frac{H_2}{H_1}\right)^{1/2} \qquad (N = RPM)$$

Depending on the ratio of the heads, the RPM increase will vary. For $H_2 = 300$ FT, $H_1 = 260$ FT, and $N_1 = 1800$. $N_2 = 1{,}934$ RPM (7.4% higher). For $H_2 = 250$ FT, $H_1 = 110$ FT, and $N_1 = 1{,}800$. $N_2 = 2{,}102$ RPM (16.7% higher).

For operation with a full tank, start at point 1. At 1a, switch to point 2a (N_2). Point 3 will be reached when the tank will be almost empty.

The average capacity will be almost constant (Q′), and the length of the operation will be a function of it. By means of variable speed, we can maintain an overall constant capacity.

■ EXAMPLE 19:2

Discuss in general terms variable demands for a fixed system (fig. 19.02).

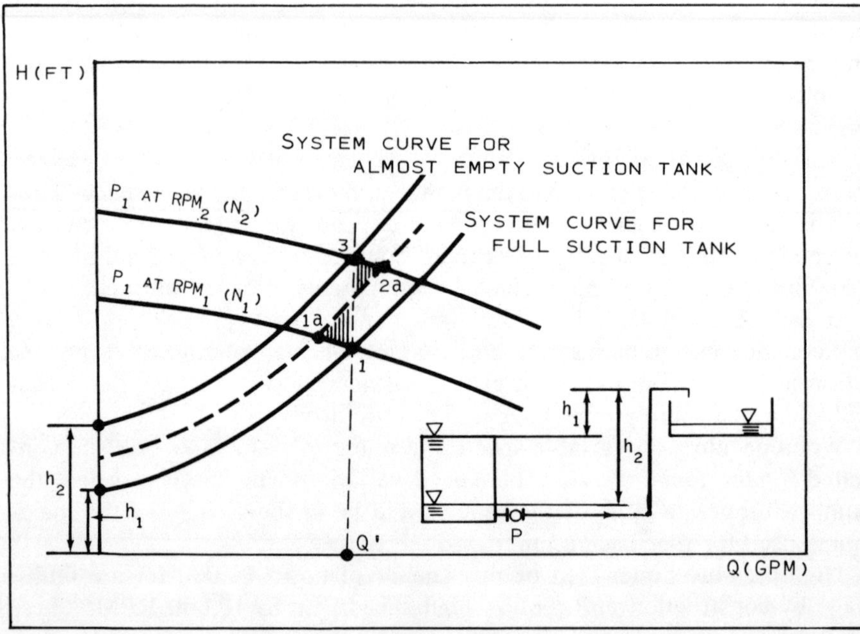

FIG. 19.01

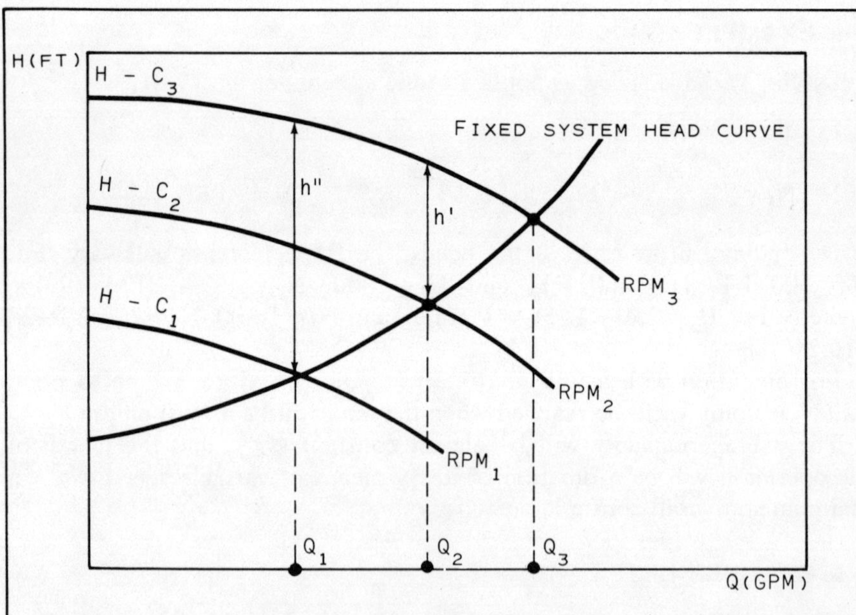

FIG. 19.02

Normally, the variations in demand control the pump speed by changing it to suit. If $H-C_2$ represents the pump curve at rated 1,800 RPM, the speed will increase for Q_3 and decrease for Q_1.

The same effects can be achieved by throttling, running the pump at constant speed. $H-C_3$ would be the selected pump curve. Q_2 and Q_1 can be obtained by throttling energies equivalent to h' and h''.

An analysis will be required for this case in order to determine the optimum solution. If Q_2 and Q_1 are required for short intervals, and Q_3 is predominant, the fixed speed system will be preferred. In case Q_2 and Q_1 are predominant, a variable-speed system will be necessary, unless one selects pumps in parallel or series to produce the required points of operation.

Again, many factors have to be considered: frequency of operation, requirement for a spare unit, space available, controls, operational skills, etc. Sometimes, the best solution from a theoretical point of view doesn't prove to be the most practical.

20.

Economic Evaluation of a Pumping System

A pumping system consists of a number of pumps and their drivers, suction and discharge headers, pipelines, valving, instrumentation, and miscellaneous equipment, plus such secondary equipment as foundations, buildings, electrical switchgear, wiring, special controls, etc. The secondary equipment will be designed after an optimization study has established the size of the piping system, the number and size of the pumps and drivers.

Information required for an economic evaluation includes:

1. Number of pumps and their efficiencies.
2. Number of horsepower-hours a year.
3. Size and lengths of pipes.
4. Maintenance required in hours per year.
5. Cost of spare parts required each year.
6. Cost of all major equipment.
7. Installation cost.
8. Cost of power.

General Considerations.

- PUMPING SYSTEM IN A PLANT.

Since the pipelines are short, their cost, if a larger size is selected, would affect only insignificantly the overall cost. Efficient and low HP pumps will definitely contribute to the overall economic evaluation of the system. The introduction of automatic controls, although expensive, should be considered, because they will reduce the operating costs.

- PUMPING SYSTEMS FOR REMOTE TRANSFERS.

In this case, the pipelines are long, and they represent the major part of the total investment. The sizing of a long pipeline has to be done very carefully. Negligence in this respect can be very costly. One should definitely be equipped with latest information on equipment, labor, and power costs.

▪ ANNUAL COST COMPUTATION

To compare two or more alternatives for overall cost, one can use various cost comparison formulae found in engineering economics textbooks. For the examples to follow we will use the exact method:

$$\text{Annual Cost} = (P-L) \times \frac{i}{(1+i)^n - 1} + Pi + OC \quad \text{.............. 20–01}$$

where:

P = initial cost, in \$
L = salvage value of equipment after n years
n = life of asset, in years
i = interest rate (in percent, fraction)
OC = annual operating costs (maintenance, operation, spare parts), in \$.

When n is large (10, 15 years, or more), L can be neglected. Then equation 20–01 changes into:

$$\frac{Pi}{(1+i)^n - 1} + Pi + OC \quad \text{.............. 20-02}$$

The expression $i/[(1+i)^n - 1]$ is very small for a large n and a small i, and can be neglected. Then the annual cost is represented by the formula:

$$Pi + OC \quad \text{.............................. 20-03}$$

▪ EXAMPLE 20.1

A product (5 cs, sp.gr. 0.7) has to be transferred from a process vessel in a refinery (pressure 160 PSIG) to another vessel (pressure 170 PSIG) through a single line at a rate of 3000 GPM. Operation is 24 hours a day. Service is critical. Distance between process vessels is 1,000 FT.

Select the number of pumps, size the suction and discharge lines. Assume 200 equivalent FT for the suction line and 1500 equivalent FT for the discharge line. Other requirements are: discharge line flow velocity limitation: 12 fps, 20 year life, 10% interest.

Solution: Since the service is critical, one must consider one spare pump. In order to counter a power failure, the spare pump could have either a diesel engine or turbine as driver. The alternatives could be: one operating pump and one spare, or two operating pumps and one spare. One can assume that space would be made available for whatever the optimum solution would require in equipment.

Generally speaking, a single pump performing is of advantage, as compared to two pumps operating in parallel. A group of pumps, in this case two operating and one spare, will provide flexibility.

The flow velocity limitation results in a 10″ line minimum for the discharge line. On the suction side, due to high pressure in the suction vessel, there wouldn't be any problems with NPSHA and NPSHR. On the other hand, due to hydraulic reasons, the suction velocity should remain below 5 fps.

This system falls under the category, "Short Lines", as defined above. Therefore, a 12″ discharge line (velocity for 3,000 GPM=8.6 fps) and a suction velocity of about 3 fps would be preferred.

The economic evaluation will concentrate on two pumps (one operating, one spare) with a 10″ discharge line vs a 12″ discharge line and a suction pipe of a size corresponding to 5 fps and 3 fps.

Suction line:
16″ line—4.78 fps, friction loss for 200 FT=1.00 FT
20″ line—3.00 fps, friction loss for 200 FT=0.33 FT
Discharge line:
10″ line—12.2 fps, friction loss for 1,500 FT=67 FT
12″ line— 8.6 fps, friction loss for 1,500 FT=27 FT
Total head:
Static Head differential=0 FT
Pressure differential=10 psi=33 FT
Friction loss in 10″ line=67 FT
Total Head=100 FT
Pressure drop for the 20″ suction line is negligible.
Friction loss in a 12″ line=27 FT
Total head=60 FT
Total BHP for an assumed pump efficiency of 80% is:
BHP for 10″ line=3,000×100×0.7/(3,960×0.8)=66 (refer to 8–02)
BHP for 12″ line=3,000× 60×0.7/(3,960×0.8)=40
Size of the drivers is 75 HP and 50 HP respectively.
Cost of pipe with valves installed is $60,000 for the 10″ line, and $72,000 for the 12″ line.
Cost of two pumps and drivers installed:
75 HP: 2×7,500=$15,000
50 HP: 2×5,500=$11,000
Power consumption and cost (1 KW=1.34 HP):
10″ line—24 hrs×365^{d}×66$^{(BHP)}$=578,160 HP-HRS/YR=
432,000 KW-HRS/year
12″ line—24 hrs×365^{d}×40$^{(BHP)}$=350,400 HP-HRS/YR=
262,000 KW-HRS/year
At $0.03/KWHR, the power costs for the lines would be:
10″ line—432,000×0.03=$12,960/year

12″ line—262,000×0.03=$ 7,860/year

Operating cost is assumed equal for both cases:

Operation and maintenance at 8 hours a day=2,900 hours/year at a rate of $20/hour complete hourly cost (salary, benefits, pension, etc); the total=2,900×20=$58,000/year.

SUMMARY OF COSTS:

LINE SIZE	10″	12″
Piping, valves	$60,000	$72,000
Pumps, drivers	$15,000	$11,000
P	$75,000	$83,000
Power	$12,960/yr	$ 7,860/yr
Oper. and maint.	$58,000/yr	$58,000/yr
OC	$71,000/yr	$66,000/yr
Total yearly cost:	$78,500	$74,300 (per 20-03)

In conclusion, a 12″ discharge line and a 20″ suction line, with 50 HP motors represent a more economical solution.

■ EXAMPLE 20.2

Refer to example 20:1 and use all data in it, except for the following changes: operation=5 hours a day, operation and maintenance=2 hrs day, and service is not critical.

SUMMARY OF COSTS:

LINE SIZE	10″	12″
P	$75,000	$83,000
Power: 5/24x12,960=	$ 2,700/yr	
5/24x 7,860=		$ 1,640/yr
Oper. and maint.: 2/8 of 58,000	$14,500/yr	$14,500/yr
OC	$17,200/yr	$16,140/yr
Total yearly cost:	$24,700/yr	$24,440/yr (per 20-03)

An exacter computation, using formula 20-02, would yield $26,000/yr and $25,900/yr, respectively.

In conclusion, for all practical purposes, the systems are identical. Therefore, the selection would be again a 12″ discharge line and 50 Hp motors, considering future capacity increase.

Note that equipment costs and labor and maintenance costs were approximated for the purpose of illustrating the method. Although approximate, they apply to both cases. Thus the application engineer is provided with a tool which will indicate the path to choose. In case of a tie, one could, as indicated above, consider future expansion and changes. This consideration, if proven true, would be a better investment.

■ EXAMPLE 20.3

Investigate the case per example 20:1 for 3 pumps (2 operating and one spare).

In this case, each pump will be sized for 1,500 GPM (3,000/2) and the same heads. Assuming equal efficiencies, we obtain 33 BHP per pump for the 10″ line, and 20 BHP per pump for the 12″ line. Motor sizes are 40 and 25 HP. Maintenance and operating time for three pumps will be 10% higher.

Cost of three pumps installed is as follows:
10″ line—3×40 HP—$4,500 each×3=$13,500
12″ line—3×25 HP—$2,500 each×3=$ 7,500
Thus initial cost **P**:
10″ line piping—$60,000+pumps and drivers, $13,500=$73,500
12″ line piping—$72,000+pumps and drivers, $ 7,500=$79,500
Power and operating cost (OC):
10″ line power—$12,960+maintenance, $63,800=$76,860/yr
12″ line power—$ 7,860+maintenance, $63,800=$71,660/yr
With formula 20-02:
10″ line=$85,400/yr; 12″ line=$81,000.

In conclusion, a three-pump, 12″ system is less expensive than a 10″ system, but more expensive than a two-pump system with a 12″ discharge line. (Compare with results in 20:1). The analysis, therefore, indicates a definite preference for a two-pump system with a 12″ discharge line.

■ EXAMPLE 20.4

Crude oil (sp.gr. 0.85, viscosity 10 CS) has to be transferred from one storage facility to another, 40 miles away, through a single pipeline. The system to be optimized comprises the pumping station (number and size of pumps), and the size of the pipeline. Consider a 16 hour a day operation, a transfer rate of 30,000 GPM, and disregard static head differentials.

Solution: The optimization will be done based on total annual cost. This considers the interest on the first cost (pumping station and pipeline), assumed at 10% a year, and a life of 25 years. Salvage value is 0, and operating and maintenance cost are on an annual basis (salaries, fringe benefits, spare parts, etc.)

The important factor in this case is the pipe size and its wall thickness. Bear in mind that pipe cost increases in direct proportion to its weight. In this pipe size range (30″-46″), for a length of 40 miles, ⅛″ added to the thickness represents an increase in the pipe cost by about $2,400,000 (Weight increase for each ⅛≅40 lbs/ft≅4,000 tons for 40 miles. Cost at $600/ton≈$2,400,000). This increase alone would result in $240,000 annual cost at the interest rate of 10%.

It follows that pipe diameter and thickness should be very carefully investigated. ANSI code B-31.4 specifies the minimum thickness as t (in.)=P×D/2S; where P=operating pressure in psig; D=pipe ID in inches; S=allowable stress (psi), per same code, depending on pipe materials. For practical purposes (corrosion protection, surge pressures, etc.), one selects a higher value than the one calculated per above formula. For this example, we will consider a thickness of 0.5″.

It follows that for t=0.5″, and a diameter of approximately 40″, the allowable pressure P would amount to 1,000 psig, allowing a value of 40,000 psi for S.

With t=0.5″ and P=1,000 psig, one can reduce the number of variables in the code formula for thickness to two (D and S), where S=1,000 ×D/(0.5×2)=1,000 D.

With a tentatively assumed velocity of 10 fps, we can obtain with formula 1-22:

$V=Q/(2.448\times ID^2)$, ID=35″, and OD=36″.

Then S=1,000×35=35,000 psi. With this value, one can select a suitable pipe material with a stress value allowable by the code. Specification API 5L×52 indicates a minimum yield strength of 52,000 psi and an allowable stress of 37,450 psi.

Summary (tentative): Pipe thickness=0.5″, pipe classification=API 5L×52, and working pressure=800-1,000 psig, to be determined further, upon selection of pipe size.

Range of pipe sizes to be considered is the smallest size corresponding to an operating pressure of 1,000 psig and the largest size depending on annual cost optimization.

Pumps: Based on the pressures computed, two or three pumps could be selected to operate in series, one of them acting as a spare. One should bear in mind that pumps of this size require a very high NPSH, 100 FT of water and more. When taking suction from tanks, one will not be able to satisfy such requirement. Therefore, pumps with lower NPSHR will have to be used to provide the same capacity per pump, but at a pressure of only about 150-200 FT, sufficient to overcome the friction losses and make the required head available at the entrance of the large line pumps. The tentative arrangement is shown schematically on fig. 20.01.

The NPSH booster pumps have to be selected for at least half the nominal flowrate in order to lower their NPSHR. Two 15,000 GPM pumps with a third one as a spare, could be selected with a NPSHR of no more than 20 to 25 FT. The size and number of these pumps will be included in the economic analysis. The total cost of the pumping station, comprising all the pumps and drivers, is a function of the total Horsepower. For the operation, refer to fig. 20.02.

Determination of Range of Pipe Sizes: Allowing for a maximum of

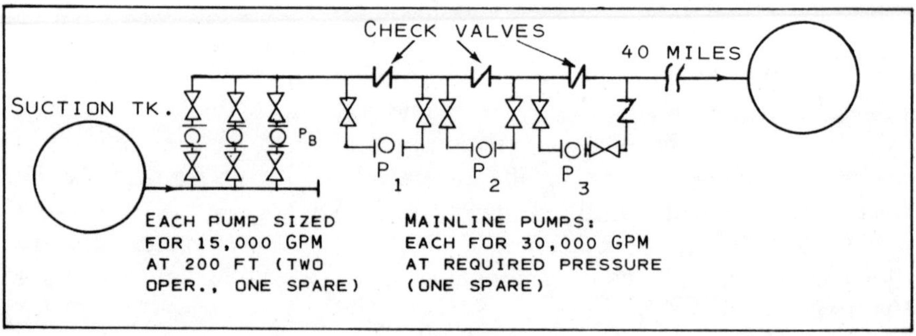

FIG. 20.01

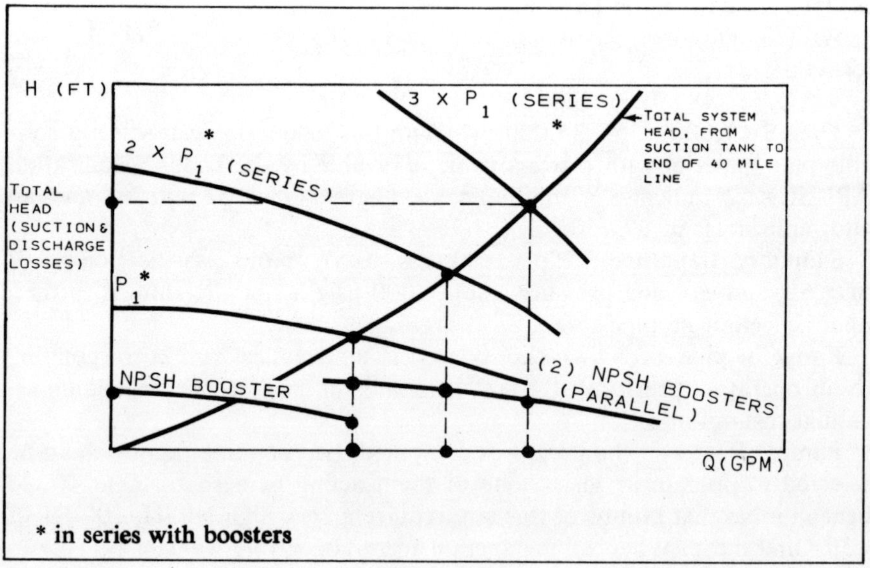

FIG. 20.02

1,000 psig pressure for the main line, one can tentatively establish the minimum pipe size for this range:

1,000 psi=1,000×2.31/0.85=2,720 FT of product=friction loss; 40 miles=212,000 FT (1 mile=5,280 FT).

Assuming a friction factor of 0.017, one can compute the ID″:

$$H_f = f\frac{L}{ID^5}\times\frac{Q^2}{32.18};$$

$$ID^5 = \frac{0.017\times212{,}000\times30{,}000^2}{2720\times32.18}$$

ID=33″

Pipeline sizes to be investigated with a 0.5″ wall thickness:

OD″	ID″
34	33
36	35
38	37
40	39
42	41
44	43

In order to compute the friction losses, one can compute an average Reynold's number and friction factor for an ID=37″:

RE=3,162×30,000/(37×10 CS)=256,378

$f=0.184/RE^{0.2}=0.015$

Compute the friction losses for all six cases, using the formula:

$H_f=f\frac{L}{ID^5}\times\frac{Q^2}{32.18}$, and tabulate the results:

PIPE SIZE		FRICTION LOSS (H_f)		FLOW VELOCITY
OD″	ID″	FEET	PSI	FPS
34	33	2272	835	11.25
36	35	1695	623	10
38	37	1283	472	8.95
40	39	986	362	8.06
42	41	768	282	7.29
44	43	605	222	6.62

Horsepower required for mainline pumps:

BHP=H×Q×sp.g/(3960×pump eff.), where H is in feet, Q is in GPM, and efficiency is in %, fraction (assumed 0.88 for this size pump).

BHP=H×30,000×0.85/(3960×0.88)=7.31×H

Compute H, BHP, number of pumps and motors, and tabulate the results:

OD″	ID″	HEAD (ft)	BHP; No OF PUMPS OPERATING	No OF MOTORS; HP
34	33	2,272	16,608; (3) at 757 ft each	(3), 6,000 each
36	35	1,695	12,390; (2) at 850 ft each	(2), 7,000 each
38	37	1,283	9,378; (2) at 650 ft each	(2), 5,000 each
40	39	986	7,207; (2) at 500 ft each	(2), 4,000 each
42	41	768	5,614; (1) at 770 ft each	(1), 6,000 each
44	43	605	4,423; (1) at 650 ft each	(1), 5,000 each

The horsepower required for NPSH booster pumps (0.82=assumed efficiency):

BHP=H×Q×sp.gr./(3,960×0.82)=200×15,000×0.85/(3,960×0.82)=785 per pump.

For two operating and one spare: 2×785+1×785 we have 1,570 operating BHP. Motor HP for each pump=1,000 HP.

The installed horsepower (operating and spares—one spare each for main-line pumps and booster pumps) is tabulated below:

OD″	TOTAL No OF PUMPS	TOTAL HP INSTALLED (INCL. SPARES)
34	3 +1 +2 +1 = 7	4x6,000 + 3x1,000 = 27,000
36	2 +1 +2 +1 = 6	3x7,000 + 3x1,000 = 24,000
38	2 +1 +2 +1 = 6	3x5,000 + 3x1,000 = 18,000
40	2 +1 +2 +1 = 6	3x4,000 + 3x1,000 = 15,000
42	1 +1 +2 +1 = 5	2x6,000 + 3x1,000 = 15,000
44	1 +1 +2 +1 = 5	2x5,000 + 3x1,000 = 13,000

■ COSTS

Pipeline: The installed cost of the pipeline is based on the costs of the material and labor plus 25% for contractor's profit and overhead. Material cost is based on $500 per long ton for pipe sizes 34″, 36″, and 38″ OD, and $600 per long ton for pipe sizes 40″, 42″ and 44″ OD.

Cost of labor varies with the circumference of the pipe. Assume $19/foot for a 34″ OD line, and prorate for the larger sizes.

The cost of materials and labor are tabulated below:

OD″	ID″	WEIGHT (LBS/FT)	TONS/40 MILES*	$/40 MILES	LABOR ($)
34	33	178 (½″TH.)	appr. 17,100	appr. 8,550,000	4,000,000
36	35	189 ″	″ 18,200	″ 9,100,000	4,235,000
38	37	200 ″	″ 19,200	″ 9,600,000	4,470,000
40	39	211 ″	″ 20,250	″ 12,150,000	4,705,000
42	41	220 ″	″ 21,100	″ 12,660,000	4,940,000
44	43	230 ″	″ 22,165	″ 13,300,000	5,175,000

*includes appr. 2% for breakage, rejects, and waste

Installed costs, including materials, labor, and contractor's fee at 25%, are tabulated below:

OD		INSTALLED COST
34″	12,550,000 x 1.25 =	$15,688,000
36″	13,355,000 x 1.25 =	$16,668,000
38″	14,070,000 x 1.25 =	$17,587,000
40″	16,855,000 x 1.25 =	$21,068,000
42″	17,600,000 x 1.25 =	$22,000,000
44″	18,475,000 x 1.25 =	$23,094,000

Installed cost of the pumping station, based on $85/HP for 20,000 HP and over, and $100/HP for less than 20,000 HP, are tabulated below:

OD"	TOT. INSTALLED HP	INSTALLED COST OF PUMPING STATION
34	27,000	\$2,300,000
36	24,000	\$2,040,000
38	18,000	\$1,800,000
40	15,000	\$1,500,000
42	15,000	\$1,500,000
44	13,000	\$1,300,000

The costs include cost of pumps and drivers, internal electrical equipment, electrical wiring, structure, lighting, controls, instrumentation, valving, piping manifolds, labor cost for construction of the pumping station, installation of all the above equipment, and taxes.

The power costs are based on \$0.075 per KWHR for up to 30 million KWHRS/YR, \$0.065 from 30 million to 50 million KWHRS/YR, and \$0.055 from 50 million to 100 million KWHRS/YR.

The operating horsepower and power costs are tabulated below:

OD"	BOOSTER PUMPS HP		MAINLINE PUMPS HP	TOT. BHP	TOT. KW*	KW/YR†	\$/YR
34	1570	+	16,608	18,178	13,550	$79{,}132\times10^3$	$4{,}352\times10^3$
36	1570	+	12,390	13,960	10,406	$60{,}771\times10^3$	$3{,}342\times10^3$
38	1570	+	9,378	10,948	8,161	$47{,}660\times10^3$	$3{,}097\times10^3$
40	1570	+	7,207	8,777	6,542	$38{,}205\times10^3$	$2{,}483\times10^3$
42	1570	+	5,614	7,184	5,355	$31{,}273\times10^3$	$2{,}032\times10^3$
44	1570	+	4,423	5,993	4,467	$26{,}087\times10^3$	$1{,}956\times10^3$

*1 KW = 1.3415 HP, † $16^h\times365^d$ = 5,840 hrs/yr.

The maintenance cost, including salaries, spare parts, maintenance, etc., is approximately \$200,000/yr for each case. The first cost and operating cost are tabulated below:

FIRST COST P (\$)		OPERATING COST (OC)*, \$
(MAT'L & LABOR) × 1.25 + PUMPING STATION		POWER & MAINTENANCE
34"	17,988,000	4,552,000
36"	18,708,000	3,542,000
38"	19,387,000	3,297,000
40"	22,568,000	2,683,000
42"	23,500,000	2,232,000
44"	24,394,000	2,156,000

*Operating HP yearly cost plus maintenance and operating cost.

The annual cost (AC) is given by:

$$AC=Pi\left[\frac{1}{(1+i)^n-1}+1\right]+OC$$

For i=10%, n=25 years, $1/(1.1^{25}-1)=0.1$, and AC=0.11 P+OC

The annual costs are tabulated below:

OD″	0.11P ($/YR)	OC ($/YR)	TOTAL YEARLY COST ($/YR)
34	1,979,000	4,552,000	6,531,000
36	2,058,000	3,542,000	5,600,000
38	2,133,000	3,297,000	5,430,000
40	2,482,000	2,683,000	5,165,000
42	2,585,000	2,232,000	*4,817,000
44	2,683,000	2,156,000	4,839,000

*Minimum.

Conclusion: Based on above costs and rates, a 42″ line (½″ wall thickness) with one mainline pump operating and one spare would be the optimum solution. Two NPSHA booster pumps operating with one spare will be required for any of the discussed cases. Assuming that all costs used above are actual, the 44″ line follows as second best.

It should be borne in mind that the selected optimum size is a function of the interest rate and of the number of years established for the life of the equipment. A temporary installation with a substantial salvage value of the installed equipment would yield a different result. A careful study is required to determine the going interest, as well as the labor rates.

Operating a 42″ line per the above will produce lesser stress:

$S = PSI \times ID''/(\text{wall thickness, in.} \times 2) = 282 \times 41/(½ \times 2) = 11{,}560$ PSI.

The code will permit the use of pipe grade A 53 which is less expensive than 5L×52, selected tentatively.

■ FINAL SUMMARY:

Although not all, sufficient number of factors were taken into consideration to determine the optimum size pipe diameter. Inclusion of all factors would in all probability not have changed the obtained result, providing the material and labor costs are actual. We can now specify the pumps:

$RE = 3162 \times 30{,}000/(41 \times 10) = 231{,}365$

$f = 0.184/RE^{0.2} = 0.0155$

$H_f = 0.0155 \times 40 \times 5{,}280 \times 30{,}000^2/(41^5 \times 32.18) = 790$ FT.

Note that 790 FT is the head loss in the 40-mile discharge line for a flowrate of 30,000 GPM. At this point, we have to take into consideration the head developed by the NPSH booster pumps. We had assumed a head of 200 FT.

Assuming that the head loss in the suction line between the suction storage tanks and the mainline pumps amounts to 80 FT, the booster pumps will contribute to the total system by providing a head of 200−80 =120 FT at the mainline pump suction. The head of 120 feet is a required condition for the NPSH of the mainline pumps.

The total (suction and discharge) head of the system is then 80+ 790=870 FT. Of these 870 FT, 200 will be supplied by the booster pumps and 670 FT by the mainline pump (Refer to note in chapter 8, at definition of NPSHR).

Number and Rating of Pumps:

Booster Pumps: three pumps (two operating, one spare)—each at 15,000 GPM, 200 FT, and 785 BHP (1,000 HP motor).

Mainline Pumps: two pumps (one operating, one spare)—each at 30,000 GPM, 670 FT, and 4,950 BHP (6,000 HP motor). Layout per fig. 20.01 (2 pumps). Characteristics per fig. 20.03.

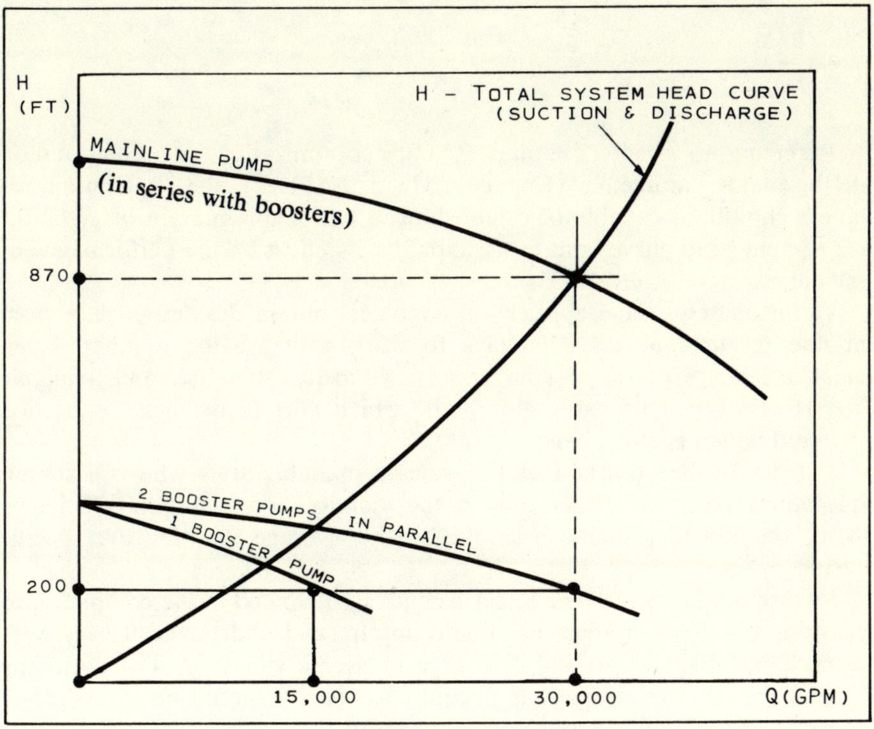

FIG. 20.03

The reader is invited to perform an analysis for a pipe thickness of ⅜″, instead of ½″, and compare the results.

Generally speaking, thickness of long pipelines vary between ¼″–¾″, depending on terrain, soil conditions, etc. The determination of the thickness for long lines requires an elaborate study, beyond the scope of this book.

The only purpose of this example is to indicate to the application engineer the major steps he has to follow while performing an optimization study. The obtained results based on approximate costs should not be considered final.

21.

General Guidelines Covering the Selection and Control of Centrifugal Pumps

Referring to chapters 8 and 9, upon optimization and computation of the major parameters (Capacity, Head, NSPHA), the application engineer should be capable of establishing a reasonable margin of error for the system-head curve, and of anticipating its effect on the performance of the curve.

Unfortunately, some application engineers obtain the design data from another group, and all they have to work with are the required head, capacity, temperature, specific gravity, viscosity of liquid, and available NPSH. They will probably also know which part of the head is friction loss and which is static head.

A form is filled out and sent to several manufacturers who will submit quotations on pumps operating in the vicinity of the highest efficiency. Also, the impeller diameter could be the average size for the specific casing.

In case the application engineer has not participated in the computations covering the whole system, he should inquire and understand it very well, especially if the suction and discharge pressures will vary. The reasoning behind the determination of the required capacity should be investigated, as well as the type of operation.

He would be displaying initiative if he would construct, on his own, the system-head curve, superimpose it on the various curves offered, and make a sound decision, based on thorough consideration of all the factors involved and discussed in this text so far.

Precautions to be taken.

The pump operating point should not be close to the end of the offered pump curve, because, if the actual head is less than the calculated one, the pump will cavitate (operate in the break). The impeller diameter should not be the largest or the smallest one for the casing.

The largest impeller will not permit an increase of capacity and head, and the smallest one will not permit decrease of head and capacity. The average size permits the flexibility. Therefore, the motor should be sized for the maximum BHP the pump may require, corresponding to the highest capacity for the largest impeller. Thus, a future change will require only the change of the impeller.

■ EXAMPLE 21.1

Refer to fig. 21.01:

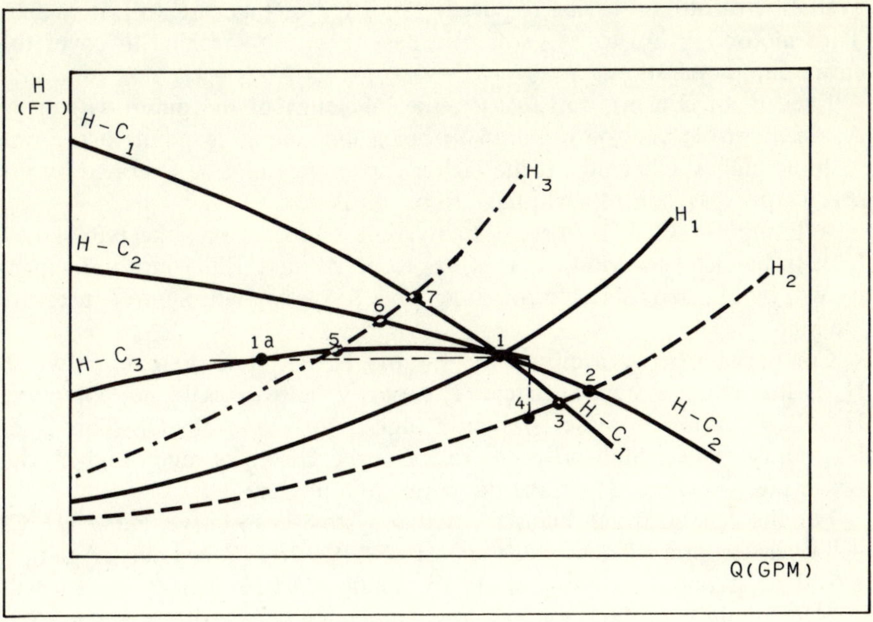

FIG. 21.01

H_1 = Computed System head curve.

H_2 = Case 1: actual system-head curve. This happens when the designer adds a little bit too much for the "unknowns".

H_3 = Case 2: actual system head curve. This happens when the designer has omitted head loss in some equipment items and underestimated pipe lengths.

1 = Computed and specified pump operating point.

Curves $H-C_1$, $H-C_2$, $H-C_3$ are pump curves offered by three different manufacturers. Note that these curves are, as in any other examples, not representative of any existing pump curves. They serve the only purpose of bringing up specifics in order to illustrate the text.

Analysis: Curve $H-C_3$ would be a poor selection. The shut-off head is lower than the operating head. This would result in an unstable operation. Point 1a on this curve has the same head as point 1. Capacity may fluctuate.

Curves $H-C_2$ and $H-C_1$ are adequate curves if one doesn't concern oneself with error margins and fluctuations, high shut-off heads. Unfortunately, disregarding these factors could result in inefficient operation. Assuming that the actual system-head curve is H_2:

- Pump $H-C_3$ would cavitate (point 4).
- Pump $H-C_2$ will operate stably, but NPSHR may be very high, and NPSHA has to be checked for the higher capacity. If not possible to make available, throttling will be required. The BHP in this case will be higher. Thus motor HP has to be verified in case it wasn't specified to cover the entire pump operation.

Even if all above is satisfactory, the efficiency of the pump will drop. Again, controls may be required to bring the operating point to the one with the highest efficiency. If the higher capacity cannot be tolerated by the specific process, controls will have to be applied.

- Pump $H-C_1$ will operate stably, but as for above, the NPSHA/R have to be checked. Motor output has to be verified. Efficiency and capacity will be affected, and controls may have to be applied. Shut-off pressure too high.

Comparison between curves $H-C_2$ and $H-C_1$: For the case in which H_2 is the actual system-head curve, capacity increase is higher for curve $H-C_2$. Pressure decrease is slightly higher for curve $H-C_1$. This is of lesser importance. Shut-off pressure for curve $H-C_1$ is much higher. Between the two, curve $H-C_2$ is the better pump curve.

For the case in which H_3 is the actual system-head curve, curve $H-C_3$ will operate more efficiently at the cost of 40% drop in capacity. Again, it would be a poor selection. (Point 5 on fig. 21.01). Curve $H-C_2$ will operate stably (point 6), but at a lower efficiency, higher head, and a 25% drop in capacity.

Curve $H-C_1$ will operate stably, at a lower than maximum efficiency (point 7), but at a higher efficiency than the one for curve $H-C_2$ in this case. It will deliver not only a higher capacity but also at a higher head. NPSHR will be lower for both cases.

It should be noted that, in the case of H_3, one cannot adjust the operation and move the operating point to 1, as with system head curve H_2. The main result of the miscalculation is lower capacity.

If the deviation is acceptable, one has to consider the specific pressures and capacities and select between curves $H-C_2$ and $H-C_3$. One would tend to select curve $H-C_2$ again because of its lower shut-off pressures and, in case capacity is a factor, to let the pump operate for a longer time.

In the case of a 24-hr a day operation and of appreciable decrease in capacity, the design mistake will have to be corrected by replacing the pump, or adding a smaller one to compensate for lost capacity. Again, careful consideration must be dedicated to the specific case before making a final selection.

In conclusion, for this case, curve $H-C_2$, as in the previous one, is the better selection.

The application engineer should inquire about minimum flowrate allowable by the manufacturer. Operators should be aware of it, so they can throttle only down to this rate in order not to overheat the liquid pumped. A pump should not be let operate long at shut-off, especially with volatile liquids. The manufacturer should be consulted about the requirement for recirculating by-pass.

On the average, the minimum flowrate equals approximately 10%–15% of the capacity at maximum efficiency.

A volute centrifugal pump is normally started at shut-off, upon which the discharge valve is opened slowly. The operating point of the pump starts at shut-off and slides downwards toward the intersection of the system curve and the pump H–C curve. It should be noted again, that axial and mixed-flow type pumps have very high power requirements at shut-off. Therefore, one starts these pumps with the discharge valve partially open.

If, during operation, a valve is closed inadvertently in the discharge line, the pump will operate at shut-off. To avoid this, controls take over and shut down the motor.

Some operations require temporary operation at shut-off pressure while, for example, a switch is made at the valved end in the pipeline in order to supply product to different consumers. In this case, shutting down the motor each time would represent a nuisance. To ease the situation, one installs a by-pass connecting the discharge side of the pump with the source of product, say a storage tank. This will allow the liquid to cool off while recirculating. In case the source is further away, one can have the by-pass connect the discharge line with the suction line. This helps only partially, but prolongs the time of dangerous heat-up. By-passes of this type are equipped with timers which are activated as soon as flow is registered through the by-pass. They would shut down the motor in a preset time of 3, 5, or 10 minutes, depending on the characteristics of the liquid pumped.

As pointed out in chapter 14, one shouldn't necessarily assume that a requirement for increase in capacity could only be met by means of adding a pump in parallel. Depending on the system curve, pumps in series may prove to be a better solution. Refer to chapter 17.

22.

Incorporation of Changes in Existing Pumping Systems

Introduction of changes into existing pumping systems requires as much diligence as that for design of new systems. Unfortunately, changes are sometimes introduced without careful investigation, and without request for professional help. In many cases, results obtained upon introduction of the changes are deplorable and costly, and require changes again in a short time.

Any change, as insignificant as it may appear, requires careful investigation of the whole system and of the effects of the change which, undoubtedly, must be taken care of, too. Many times, the required change is necessary because of a faulty and inefficient design in the first place.

Changes are also required because of expansion, change in process, etc. Under normal conditions, somehow there is not enough time to run an investigation and study the impact of the introduced change. The result, again, is a costly and inefficient operation, and downtime which in itself, depending on the vitality of the process, can cause many troubles.

The improper way to introduce a change is to instruct the engineer or operator on what to do without explaining reasons. Instead, he should be told what the problem is, what results are expected to be achieved, and let him decide what kind of changes should be introduced into the existing system. A faulty pump system can not only prove to be costly as such, but it can affect negatively a whole process.

In order to introduce a change, the application engineer must know why it is required and must be convinced that the requirement makes sense. Then he has to study the whole system and offer alternative solutions with economic analysis. The alternatives have to be discussed with the personnel responsible for the overall process. Only then can the design of the change be initiated.

A pumping system consists of different elements, and various parameters govern its operation. The following are pumping system elements which could be involved in a request for change:

1. Suction source. Vessel, storage tank, lagoon, sump, lake, sea, pipeline, etc.

2. Suction pipe. This includes the pipelines connecting the suction source with the pump, and all valves, fittings, strainers, control stations, etc. in them.

3. Pump(s) and driver(s). The pumps and the drivers and, in case of multiple pumps, the piping headers and controls.

4. Discharge pipe. This includes the pipelines connecting the pump with the consumer and all valves, fittings, equipment, control stations, etc. in them.

5. Receiving end. Vessel, storage tank, drainage system, lagoon, lake, sea, another pipeline, boiler, etc.

A request for a partial change in the existing system will affect one or more of the elements listed above. Changes affecting each one of the elements are discussed, and suitable examples are provided for illustration and clarification.

■ 1. SUCTION SOURCE

Changes affecting the suction source of a pump result in loss or increase of suction head due to changes in suction levels and pressures. For well substantiated reasons, a suction vessel may have to be reinstalled at a lower or higher level; the pressure in a suction vessel may increase or decrease; in the case of a vertical pump, the water level may drop or rise.

A change in the suction head in an existing system would result in lower or higher capacity, lower or higher head, and loss or gain in the NPSHA. The effect of the change on the head, capacity, and the NPSHA must be investigated. Only in case none of the above is affected in a negative way, causing deterioration of the pump operation, can the change be approved.

■ 1.1. CHANGE DECREASES AVAILABLE STATIC HEAD. Refer to fig. 22.01. The parameters affected are: NPSHA (decrease); capacity (decrease); power consumption (decrease); efficiency (decrease); and head (increase).

Comments: If the NPSHA turns out to be lower than the NPSHR of the pump, it has to be restored before the change is approved. For example, in the case of a vessel, the minimum level down to which pumping at the rated capacity is permissible (in order to satisfy the condition that the NPSHA must at all times remain larger than NPSHR), will have to be raised. This means that, after reaching the higher level required, the vessel will have to be refilled.

Under no conditions shall a pump be let operate with insufficient NPSHA. Corrections will have to be introduced. Among the choices are: (1) rearrangement of the pump installation, (2) providing a booster pump

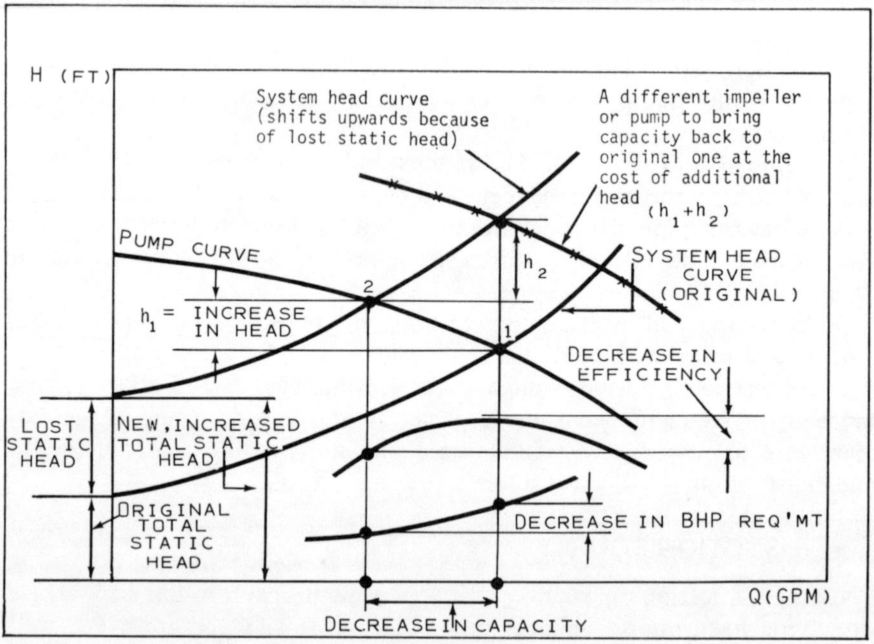

FIG. 22.01

to bring up the available head for the main pump, or (3) replacement of the pump with one for which the new NPSHA is sufficient.

If the decreased capacity is unacceptable, one has to increase it by:

1. Increase of pump speed, provided the NPSHA exceeds the NPSHR for the new operating point. This solution would represent no problem at all, in case the driver is a steam or a gas turbine, or a diesel engine. In the case of an electric motor, one has to replace it with one at the suitable speed, or with a variable-speed motor.

2. Installation of an additional pump in series to increase the capacity to the previous value. Analysis of the operation schedule is required, though.

3. In the case of a relatively short discharge line, it can be replaced with one of a larger diameter, and thus change the system head curve favorably.

The specific case may offer other solutions.

The increased head will normally not pose problems since the system is designed to operate at shut-off head. One must, though, check for minimum allowable flowrate which varies between 10-15% of the capacity at maximum efficiency.

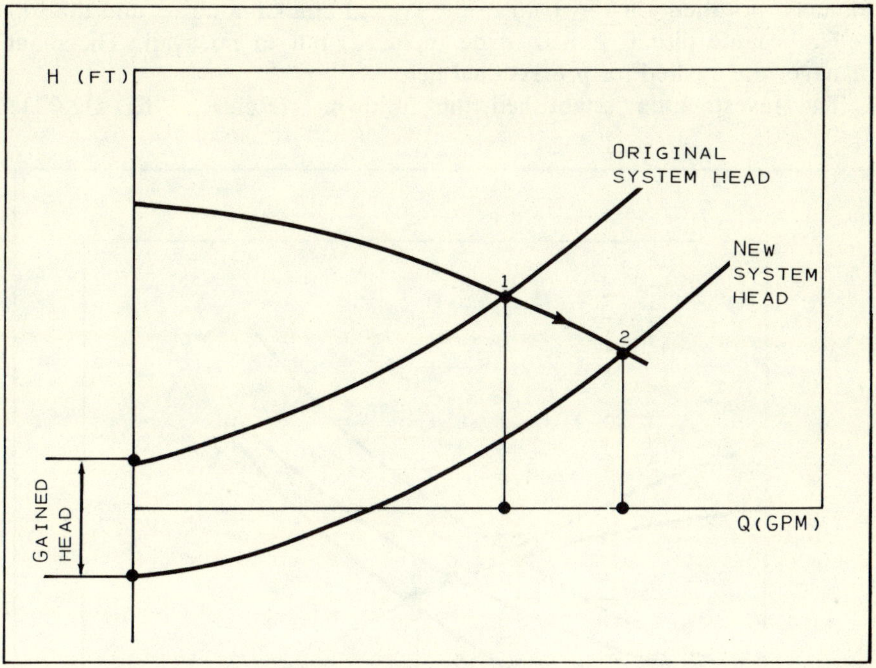

FIG. 22.02

■ 1.2. CHANGE INCREASES AVAILABLE STATIC HEAD. Refer to fig. 22.02. The parameters affected are: NPSHA (increase); capacity increase); power consumption (increase); efficiency (decrease); and head (decrease).

Comments: Point 2 may not be a stable point on the H−C curve of the pump, and it will cavitate. In this case, throttling or an automatic minimum-pressure-sustaining valve in the discharge line will help correct the condition. The correction for this case is simpler to achieve as compared to the more complicated preceding case.

■ EXAMPLE 22.1

A plant manager complained to a pump manufacturer's representative that he was not getting the desired capacity after installation of a new pump. He explained that he would accept a higher than specified capacity, but under no circumstances a lower one.

The representative suggested an increase in the suction head which is theoretically correct. The design pressure in the suction vessel, being above atmospheric, was controlable. The advice was followed. The result was that the capacity did indeed increase, but the pump started cavitating. Unfortunately, no attention was paid to the typical crackling noise, and the impeller became pitted. It had to be replaced, but to no avail. The plant manager then called for professional help.

The investigation established the following (Refer to fig. 22.03):

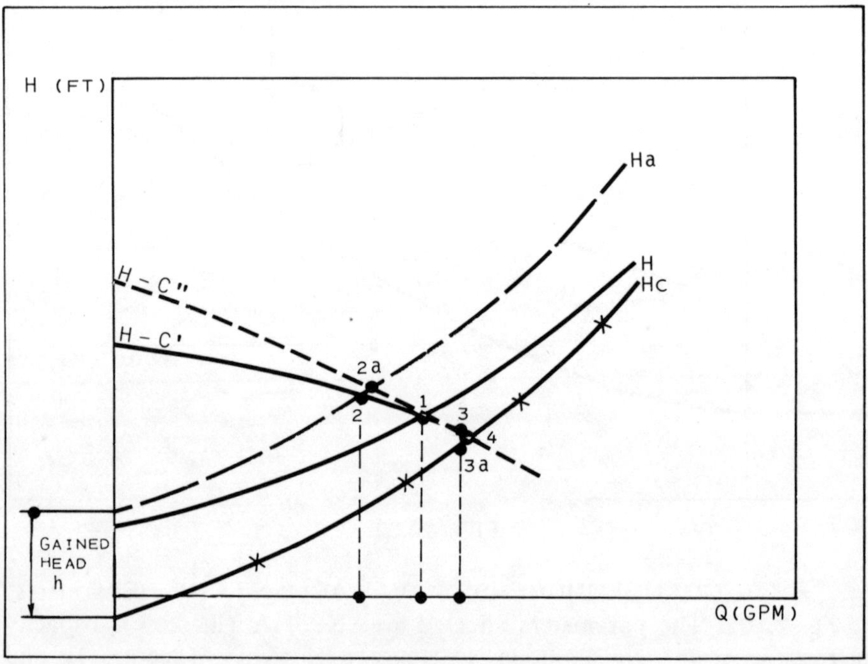

FIG. 22.03

H: the computed system-head curve; 1: the specified operating point; H−C′−3: characteristic curve of the purchased pump; H_a: actual system head curve upon installation; 2: actual operating point upon installation of the pump. (Capacity about 80% of specified); H_c: corrected system head curve by means of increasing the pressure in the suction vessel to achieve a gain in suction head corresponding to h; and 3_a: since pump curve ends at point 3, the pump operates in the break (cavitates). Capacity by now is about 8% higher than specified.

Conclusion: The pump curve is not adequate to take into consideration deviations. The pump H−C′ was replaced with pump H−C″. Operating point 4, with corrected head-capacity system, is stable. Throttling could bring the capacity exactly to the specified one. In all probability, the efficiency at point 1 would be higher than at point 4, and the loss in energy during throttling would be compensated by the increase in efficiency.

■ EXAMPLE 22.2

A pump in a plant taking suction from a vessel situated above its centerline, is, because of physical requirements, disconnected from that vessel, and connected to another suction vessel, located two plant levels lower than the centerline of the pump. The product, temperature, and pressure in the second vessel are identical to the one in the first vessel.

Aware of the fact that, due to change, the operating point would move toward lesser capacity (which in this case was acceptable), and that there would be no problems with NPSHA because of the high pressure in the suction vessel, the change was executed without hesitation and any adjustments. The pump, in the plant operator's opinion, had operated satisfactorily before the change. The result was that the product overheated, and the pump seized. Professional help was called for, and the investigation established the following (Refer to fig. 22.04):

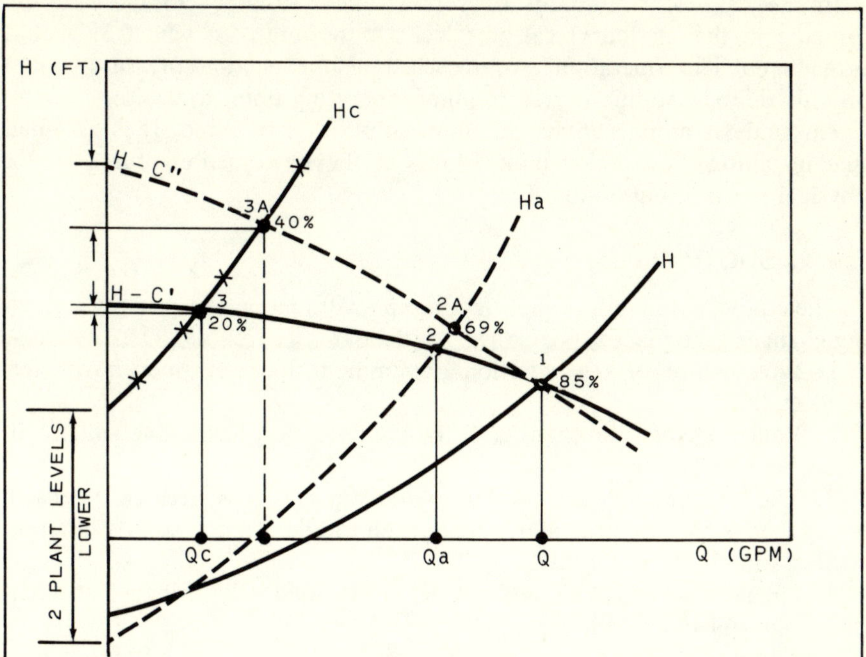

FIG. 22.04

H−C′: existing pump head capacity curve; H: computed system head curve; H_a: actual system head curve after installation; H_c: changed system head curve due to the lower suction source; H−C″: suggested more efficient pump curve; 1: specified operating point; 2: actual operating point; 3: operating point after change; and 2a and 3a: improved operating points with pump H−C″.

Comments: H−C′ is too flat a curve, to begin with. The system head curve was not computed thoroughly. The result is a 25% loss in capacity and 16% loss in efficiency. Taking suction from a vessel two plant levels lower resulted in the pump seizing due to flowrate below the minimum allowed.

One couldn't criticize the selection of the pump in general, would it have operated around point 1, as planned. A careful plotting of the system head curve had not been done.

Before introducing the change, an application engineer would have alerted the plant management of the dangers of low flowrates. A bypass would have helped, but the efficiency is very low (20%).

If H−C″ would have been selected, 3a would have been the operating point after the change. The efficiency, although twice as good as the previous one, is inadequate. H−C″ could have been tolerated with its operating point 3a, provided the operation is intermittent.

In the case of continuous operation, one has either to increase the pressure in the suction vessel or select a new pump for which 3a would provide efficient operation. An increase of the suction pressure would provide the best solution: restore pump operating point to 2.

An analysis aiming at the optimum solution is required. It is assumed that installing the existing pump down at the new suction source is, for physical reasons, out of question.

▪ 2. SUCTION PIPE

The equivalent length of the suction pipe will increase in case the following changes take place:

1. Introduction of new, additional equipment, like a strainer, flowmeter, valve etc.,
2. Rearrangement or changes in the configuration in the line which will make the line longer.
3. The distance between the pump and the suction source is increased; which is feasible when the pump is required to take suction from a different suction source, further away.
4. Partial obstruction in the line, like a throttled valve, clogged strainer, or foreign object in the line.

The consequence of the above changes will affect the NPSHA, especially if suction is taken from an atmospheric vessel; hot and volatile liquids may vaporize partially and contribute toward cavitation. In case the NPSHA is affected negatively, corrections must be made, whatever the cost. The pipeline can be replaced by a larger one, or an additional line can be run in parallel to the original one. The level in open tanks will have to be controlled so that refilling starts at higher level than previously, in order to compensate for the added losses. The pump may have to be replaced with one requiring lesser NPSHA. It is strongly recommended not to introduce any changes in a well-designed suction line without professional help.

Tapping the suction line or header, in order to feed an additional pump, must be investigated very carefully, especially that portion of the line carrying the total capacity to both pumps. Recalculation of the NPSHA is mandatory.

Throttling of a valve in the suction line should definitely be avoided. Throttling is permissible only in the discharge line.

The suction system is vital for the proper operation of the pump. Therefore, one must spend adequate time to design it, and twice as much when introducing changes. Inefficient pump operation is in most of the cases, attributed to faulty suction systems.

■ 3. THE PUMP.

The following are changes which can be introduced and which will affect the pumping performance:

1. Replacement of the pump and driver with different ones.
2. Replacement of the pump only.
3. Replacement of the driver only (size and/or type)
4. Replacement of the impeller in the existing pump.
5. Addition of one or more pumps.
6. Change in speed (RPM).

Circumstances warranting changes affecting pumping performance are:

1. Change in capacity.
2. Change in head.
3. Change of product.
4. Change in temperature and viscosity.
5. Change in the NPSHA due to changes in the suction system.
6. Change in horsepower requirements due to some of the changes above.

As we keep stressing, any introduction of changes must be preceded by a thorough investigation of the system in its entirety. The existing facilities may have been in service for a long time, and installation drawings may not be available. If available, they may not reflect the as-built condition and the small changes over the years.

The application engineer must take measurements, determine existing lengths and equipment, determine static elevations, and get familiar with the operation of the system (suction source, receiving vessels, frequency of operation, etc.). Upon compilation of all these results, the system head curve has to be computed and plotted, and combined with the existing pump curve so that conclusions can be drawn.

The circumstances warranting changes will be discussed in detail.

▪ 3.1 CHANGE IN CAPACITY

Very often, the capacity has to be increased for process development reasons. The following options can be made available:

1. Replacement of the impeller with a larger one. The result is gain in capacity and head.
2. Increase of the pump speed. This is very easily done when the driver is a turbine or an engine. In most cases, though, the driver is an electric motor. If the motor is not of the dual, or variable-speed type, it has to be replaced. Generally speaking, it is not simple and economical to increase the motor speed for the purpose of increasing the capacity of the pump.
3. Addition of a second pump in parallel, or in series. The system curve will indicate which arrangement is more efficient. (refer to chapter 17).
4. Replacement of the pump with a new one suitable for the requirement.
5. Operating the existing pump for a longer time, if feasible, in order to achieve a larger daily average.

With the increase in capacity, the head and the NPSHR will increase too. The flow velocity in the lines will be higher. The application engineer should not permit a higher discharge velocity than accepted in the industry (6-15 FT/sec, depending on the product and its temperature) Also, suction velocities should be paid attention to (2-5 fps are the acceptable values, again, depending on product and temperature).

Compromises result in inefficient operation. It follows that, if the existing system was already designed for the maximum velocities, there is no room for further capacity increases. In this case, one has either to replace the pipeline, or add a second pipe in parallel.

A decrease in capacity can be achieved by means of throttling, recirculating bypass, decrease in the RPM, if feasible, decrease in impeller diameter, or if none of these solutions proves to be practical, replacement of the pump.

▪ 3.2 CHANGE IN HEAD

Reasons for increase in pump head are:

1. Increase in capacity (discussed above).
2. Requirement for higher head for process reasons. This can be achieved by throttling (inefficient and costly), provided that the sacrifice in capacity is acceptable, and that the existing pump curve is steep in order to provide substantial change in head for a relatively small decrease in capacity. In case the capacity should not be decreased, one could replace the impeller with a larger one, or increase the pump speed. In case the capacity increase is unacceptable, one can install a bypass, connected to the suction source. For a requirement for a substantial increase in head, one could add a pump in series, or replace the existing pump with a suit-

able one. Pressure ratings of the existing elements in the system have to be checked out.

3. Introduction of new equipment in the discharge line. If the increase is within reasonable limits, one can accept it. If substantial, one will have to cope with decreased capacity and the problems connected with it, as discussed above.

■ 3.3 CHANGE OF PRODUCT.

The industry calls the change of product pumped by the same pumping system "conversion" from one product to another. The conversion may be temporary or permanent. The preparation for the conversion (purging of line, pump, elements, etc.) is beyond the scope of this book. The application engineer will require complete information about the old and new products, frequency of operation, updated layout drawings, permanency of conversion, etc.

Assuming that the suction source and consumer will continue to operate at the same pressures, the consequences of the conversion will be as follows: (1) change of capacity; (2) change of head; (3) change of efficiency; (4) change of BHP requirements; (5) cavitation, if the operating point, due to a less viscous product, shifts toward higher capacity; and (6) change in the NPSHA.

Case A. Conversion from less viscous to more viscous product.

The consequences are: decrease in capacity, increase in head, decrease in efficiency, and decrease in the NPSHA (the NPSHR will decrease also, though, slightly) due to the larger head loss. This may not necessarily be true if there is a change in the vapor pressure, such that the head loss will be compensated for.

In case all these changes are unacceptable, replacement of impeller and driver may be required. Another alternative would be to add a pump in series, in order to bring the operating point back to the same capacity as before at a higher head. This solution would be appropriate if one plans to alternate products.

Adding a pump in series always requires checking for pressure class of existing fittings. For example, if the operating pressure was 168 psig, 150 class fittings would do. But if the new pressure goes up to 260 psi and higher, one has to replace the fittings with new ones of the 300 psi class. In case the efficiency of the pump drops substantially, it would pay to replace the pump and driver (Refer to fig. 22.05).

Case B. Conversion from more viscous to less viscous product. The result is increased capacity with risk of cavitation, decrease in efficiency (assuming the previous operating point was at or around the highest efficiency), increase of NPSHR, and increase or decrease of NPSHA, provided no other changes have taken place.

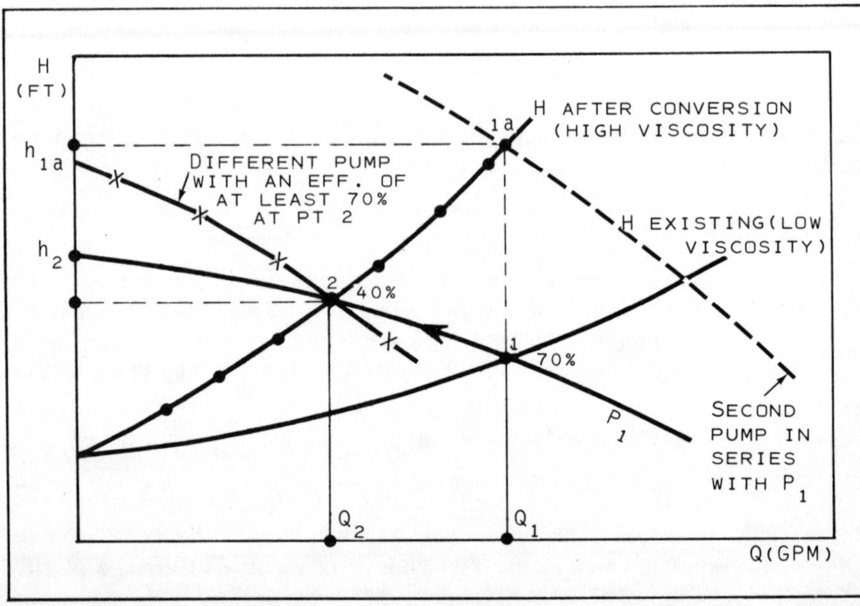

FIG. 22.05

In case the deviations are unacceptable, one can throttle and bring the operating point back to the previous one. Since, in this case, the pump will operate at its peak efficiency, the throttling loss will be compensated for. No additional changes will have to be introduced.

In case products will be alternating, a pressure control valve should be installed at the discharge of the pump maintaining minimum "non-cavitating" pressure (refer to fig. 22.06).

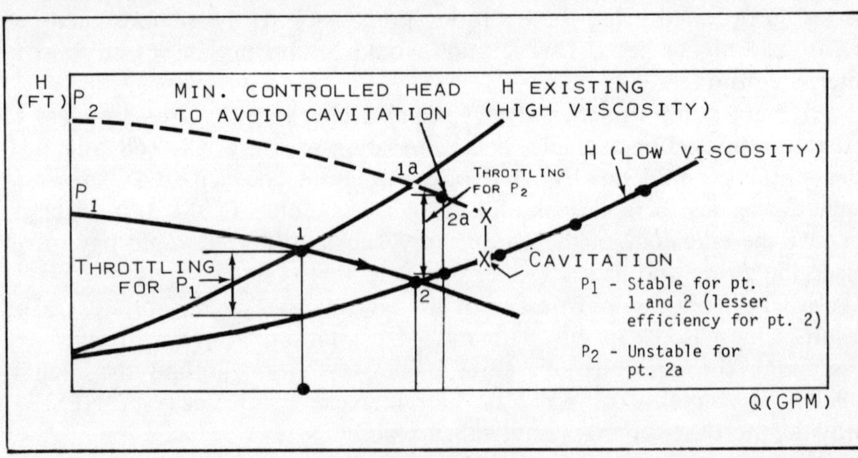

FIG. 22.06

Note that, in order to utilize a pipeline more efficiently, one practices to pump alternately two or more different products without having to purge the line, vessels, and pumps. This is possible only with products, the slight mixing of which would not matter. For example, one can pump diesel oil and then gasoline. The suction line is connected to different storage tanks, and one switches from one product to the other.

In this case, we can not say that we have a case of product "conversion". One product displaces the other. Thus, we are dealing with a sliding operating point on the pump curve. Pumping different products by means of switching can be troublesome if the system is not investigated properly.

Let's assume that the system was designed for diesel oil (specific gravity 0.87, viscosity 50 SSU), and a decision is made to switch from time to time to gasoline (sp.gr. 0.65, viscosity 20 SSU). With the discharge line full of diesel oil, we start pumping gasoline. Per fig. 22.07, the initial operating point is C.

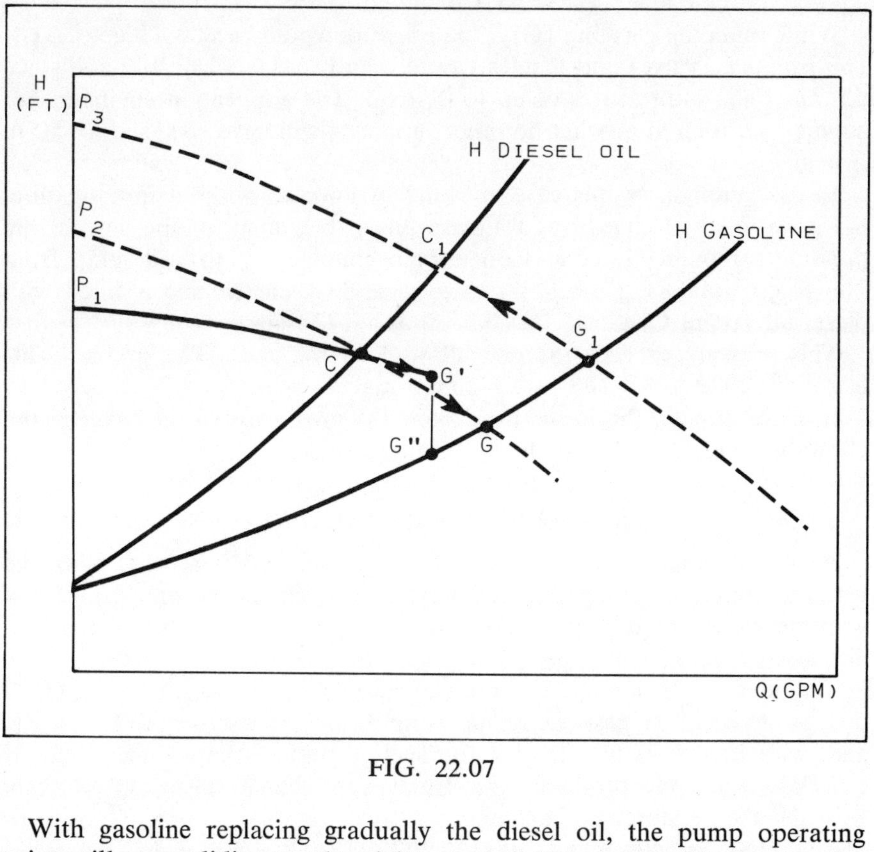

FIG. 22.07

With gasoline replacing gradually the diesel oil, the pump operating point will start sliding to the right, toward higher capacities and lower pressures. With pump P_1, cavitation will start at G′, and the operating point will establish itself at G″, when the entire line will be filled with gasoline. Therefore, P_1 is not suitable for a switch from diesel oil to gasoline.

Pump P_2, on the other hand, would be suitable. The operating point would slide toward G which eventually will become the gasoline operating point. The pump efficiency will drop, the NPSHR will increase. The NPSHA may decrease because of the higher vapor pressure of gasoline, and cavitation may occur. Therefore an investigation is mandatory.

Let us now look at a system designed initially for gasoline transfer (Pump P_3 and operating point G_1, per fig. 22.07). If we switch the service to diesel oil transfer, the operating point will slide up from G_1 to C_1. Assuming that the efficiency at G_1 is the highest, the switch will result in lower efficiency, a drop in capacity, and higher head. The NPSHR will drop, and the NSPHA will decrease.

In order to authorize a switch of this kind, the resulting pressures have to be investigated very carefully. Assume that the head at $G_1=600$ FT of liquid, and the one at $C_1=775$ FT of liquid.

While pumping gasoline (G_1), the pressure would be $600\times0.65/2.31=$ 169 psig. The valves and flanges would then be of the 150 lb class, since this class allows pressures of up to 274 psi. The application engineer, not having had in mind a switch to other products, will have specified a 150-lb system.

In case though, in this case, a switch is initiated without investigation, the pressures will change as follows: at the beginning of the switch, the pressure for point G_1 (diesel oil is being pumped now) will jump from 169 psig to $600\times0.87/2.31=226$ psig, and, when the line is filled with diesel oil (point C_1), to $775\times0.87/2.31=292$ psig.

This pressure exceeds the one allowed by the code. The pressure differential ($292-169=123$ psi) is substantial.

In case, though, the heads are lower, the pressures will not exceed the allowable ones, and the switch can be authorized.

▪ 3.4. CHANGE IN TEMPERATURE AND VISCOSITY.

A change in temperature affects vapor pressure and viscosity. With the temperature of the pumped product rising, one should be only concerned with the vapor pressure, since the viscosity decreases, and with it, the friction loss.

The effect of the increased vapor pressure could be overwhelming. Cavitation may occur. In case the temperature drops, the viscosity will increase, and, with it, the friction loss in the suction and discharge lines. This is especially valid for products, the viscosity of which varies substantially with the temperature. Some crude oils, for example, have a viscosity of 150-200 SSU at 140° F and 2,000–4,000 SSU at 80°-60° F. In such cases, pipelines have to be heated, in order to make pumping easier. With other products the variation may not be substantial, but it has to be taken into consideration when processes require changes affecting the viscosity. The system curves have to be computed for the extreme cases and superimposed on the existing pump curve for evaluation. The pump should still be capable of operating stably for the new condition. Otherwise, suitable changes, as discussed in preceding subitems, will have to be introduced.

▪ 4. THE DISCHARGE PIPE.

Changes in the discharge pipeline system can be for both physical and process reasons. Physical reasons include changing route between same originating and final points; looping the discharge pipe up, or burying the line under a road; or replacing a portion of the discharge line with a smaller size, because of unrelated construction requirements.

Process reasons include introduction of additional equipment in the discharge line (strainer, control valve, heat exchanger, flowmeter, etc.); and replacement of an old, corroded pipeline, or pipeline stretch, with a new one. Sometimes, considering future expansion (increase in capacity), and the fact that the labor cost is, for practical purposes, the same, one installs a new line one or two sizes larger.

Consequences: An increase of equivalent length can be expected for all of the above cases with the exception of the last one. The system curve will change towards decreased capacity, efficiency, and higher head.

In case these changes are unacceptable, one can replace the impeller with a larger one. Other alternatives include increase of speed, or if possible, replacement of pump, if changes are substantial.

The last case in the list above (increasing the pipe diameter) could cause cavitation due to increased capacities. Motor HP has to be checked, too. Control valves will have to be introduced to maintain stable operating conditions.

▪ 5. THE RECEIVING END OF THE SYSTEM.

The following changes can take place on the receiving end:

1. Level change of receiving vessel.
2. Substitution of a discharge into a vessel with a discharge in the open, like a lagoon, sump, etc., or vice versa.
3. Pressure change in the receiving vessel.
4. Change of number of receiving ends (splitting, or combining flows).

Consequences: The affected parameters are head, capacity, efficiency, BHP, and NPSHR. If the changes are minimal, their effect will be absorbed by the existing pump. (Refer to fig. 22.08).

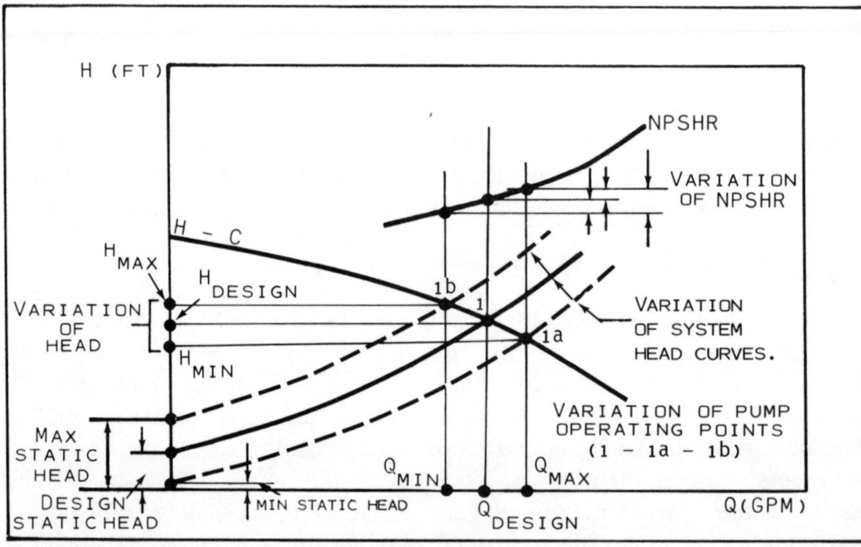

FIG. 22.08

In case the changes affect the parameters substantially, the pump will not operate stably. The discharge will have to be throttled on a continuous basis. This may represent a considerable loss in energy, therefore a pump replacement may be advisable (Refer to fig. 22.09).

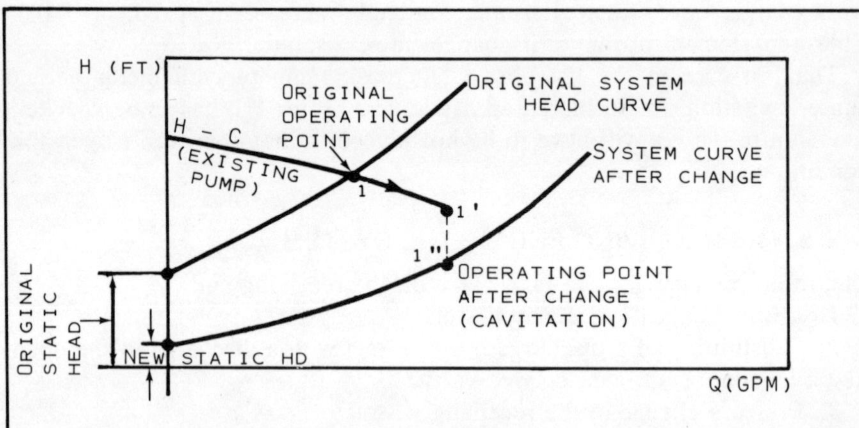

FIG. 22.09

In fig. 22.09, the changes resulted in increased capacity and in cavitation. The opposite will occur per fig. 22.10. A substantial decrease in capacity takes place. The pump may be operating below its minimum allowable capacity, and overheating may take place. The efficiency is very low, and power consumption is relatively high. The pump will have to be replaced with a suitable one (H−C_2).

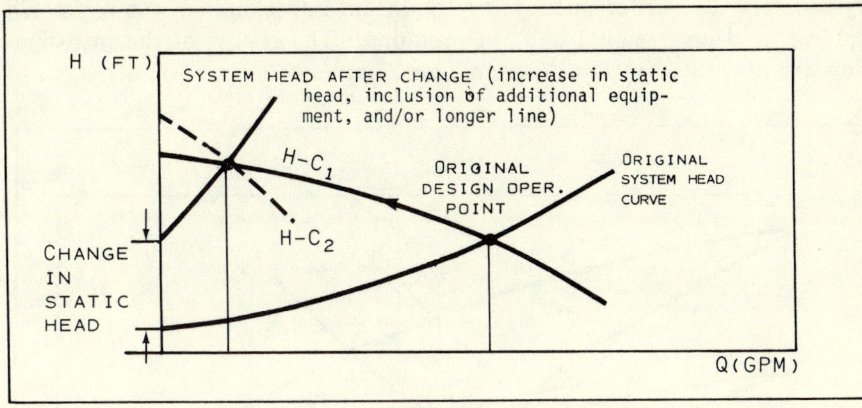

FIG. 22.10

In case of splitting flows, as shown diagrammatically on fig. 22.11, product is being transferred to two consumers instead of to one only. If the flow is simultaneous, there will be a pressure drop in the line (A−T′ for full flow requires more head than A−T′ and A−T″, since the flowrate in the branches is less than for full flow). Cavitation may occur. The remedy is control valves.

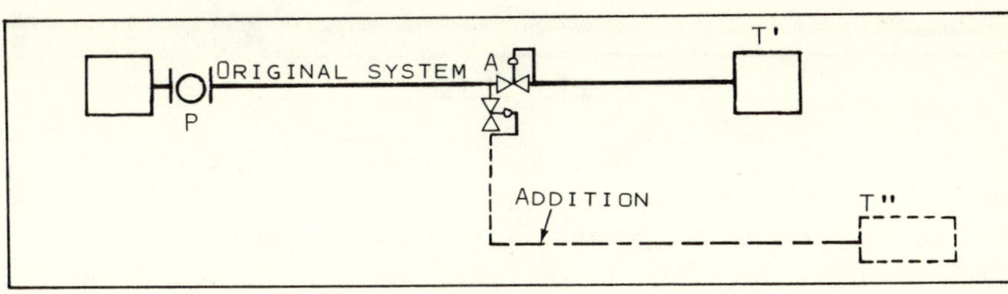

FIG. 22.11

■ EXAMPLE 22.3

An existing pump installation consisted of three identical pumps (P_1), installed in parallel; two operating; one spare. (Refer to fig. 22.12). The 12-in. discharge line was designed with future capacity increase in mind.

During the operation with two pumps, though, the pumps cavitated slightly (pt. 2). Nevertheless, the operation was continued. Occasionally, all three pumps were run, but the increase in capacity was insignificant (pt. 3).

When time came, one more identical pump was ordered and installed, in order to increase the capacity, as originally planned. Since the operation was intermittent, it was planned to run most of the time all four pumps (the existing three and the identical fourth) in order to increase the capacity substantially. Unfortunately, no significant improvement was registered (pt. 4), and professional help was required. The results of the investigation are illustrated in the diagrams and commented upon.

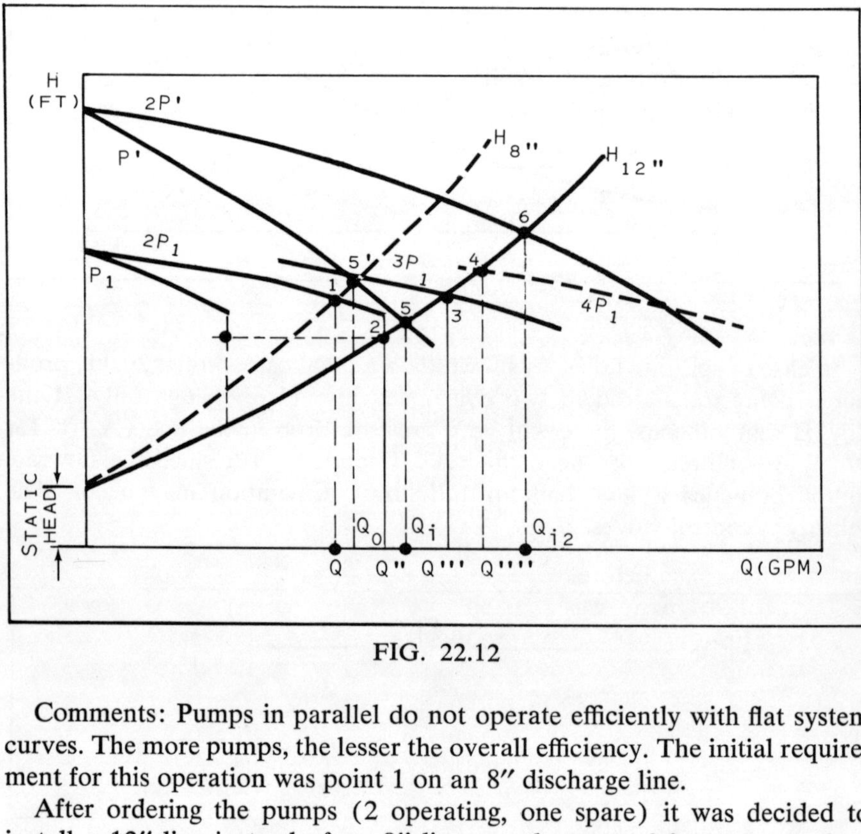

FIG. 22.12

Comments: Pumps in parallel do not operate efficiently with flat system curves. The more pumps, the lesser the overall efficiency. The initial requirement for this operation was point 1 on an 8″ discharge line.

After ordering the pumps (2 operating, one spare) it was decided to install a 12″ line instead of an 8″ line to take care of future expansion. Two pumps operated, cavitating slightly, at point 2. When the third pump was included, the operating point shifted to 3, resulting in a capacity Q^{III}, only about 15% higher. With the new, fourth pump, (operating point 4), the capacity increased by just 10%.

It is obvious that the selected pumps curves were not suitable for a 12″ line. With an 8″ line, one would have had a stable operation (point 1) had the initial size of line been retained.

This is an example of a case where scope changes are made after specifications of pumps have been prepared and not modified. The slight cavitation did not bother anyone, so a fourth (identical) pump was ordered with the hope that it would help increase the capacity substantially.

With the scope in its entirety in front of him, the application engineer should have avoided planning for so many pumps with flat curves. He should have shopped for larger pumps, like P′. Two of these should have been installed, one spare. Points 5′ and 5 would have been the respective operating points for one pump (capacities Qo, Qi).

With a requirement for intermittent increase in capacity, 2 P′ would have supplied Qi_2 (point 6), an increase of almost 30%, as compared to Qi. Four pumps cost more to install than two, even if the two are larger. Also, the larger pumps are more efficient.

Conclusion: Replace the four pumps with two larger ones.

▪ EXAMPLE 22.4

An existing pumping facility had been operating with one pump on an intermittent basis. Request for increased capacity required the installation of an additional pump. Unfortunately, no investigation whatever was initiated, and the decision was made to use an existing, larger pump, stored in the warehouse. It was felt that the two pumps, installed in parallel, would increase the capacity substantially.

Upon installation of the second pump and start up, it was found out that the larger pump was backing off the first pump, i.e., it was forcing it to operate at shut-off. Then, the second pump was let run by itself, but it cavitated, and its motor was being overloaded. Only substantial throttling could make that pump operate stably, but then the capacity wasn't too much higher than before.

The second pump was shut-off, the capacity continued to be provided by the first pump, and professional help was called for. The following facts were determined:

The first pump had been purchased upon informing a manufacturer's representative of the desired capacity to flow through a specified length of 10″ line. There was no static head to be considered. He computed the head and supplied a relatively satisfactory pump. He had not been consulted, though, about the installation of the second pump.

Fig. 22.13 shows the existing condition, before the installation of the second pump. Q′ was the specified capacity. Fig. 22.14 shows the head capacity curve of the second pump. It can develop similar capacities, but at almost double the head.

Fig. 22.15 shows the two pumps and their combined operation in parallel. Since the pump curves cannot be added for parallel operation in this case (there are no identical heads), the larger pump becomes the governing one. Thus, at all times it will be backing off the first pump and force it into shut-off operation, which is dangerous for the pump, and has, by all means, to be avoided. By itself, the second pump cavitated (point 2) because its head-capacity curve doesn't intersect the system head curve.

Conclusion: No combination with the existing first pump would result in an efficient operation. The first pump should have been selected with expansion in mind, but it was not. Therefore, the only solution is to replace the first existing pump with a new pump suitable for the required conditions.

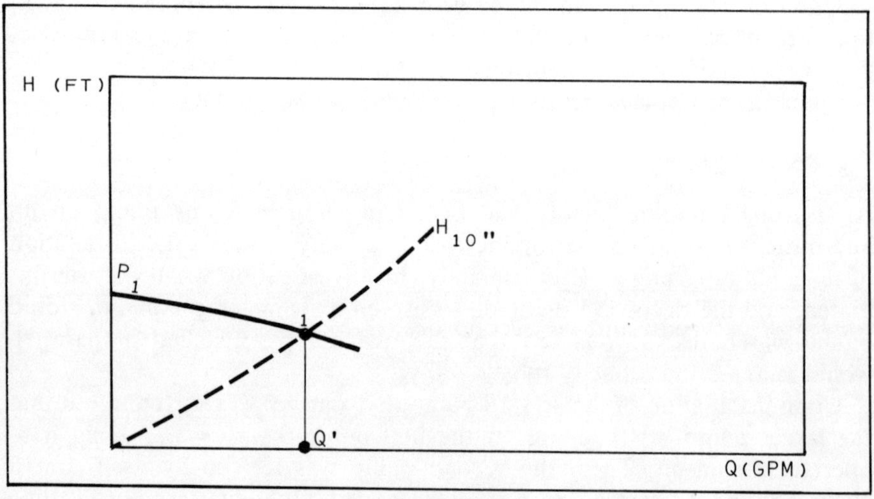

FIG. 22.13

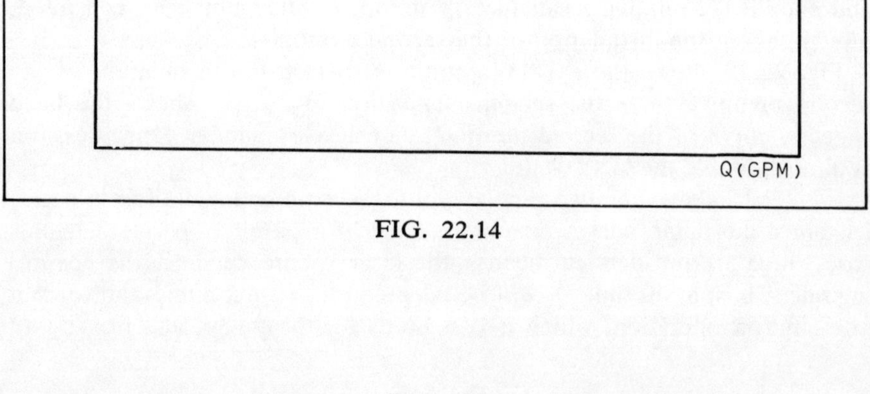

FIG. 22.14

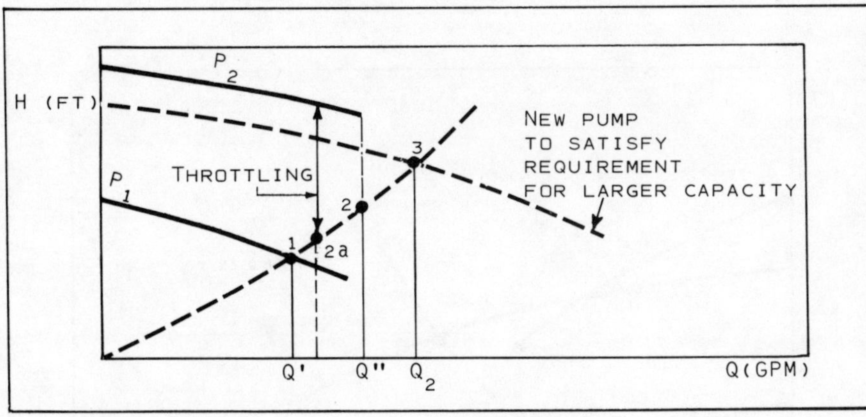

FIG. 22.15

■ EXAMPLE 22.5

A pumping installation with two pumps, arranged in parallel, used one pump for its needs, and the second one as a spare. With requirement for increased capacity, the two pumps were run in parallel, but the resulting increase in capacity was negligible.

Since the new service was supposed to be on an intermittent basis, it was decided to pipe up the two identical pumps in series which, in theory, is a sound decision. But no curve plotting was done, and the two pumps cavitated heavily when operated. The reason is obvious from fig. 22.16.

Conclusion: A new and larger pump has to be selected for the required new, increased capacity. In case the original capacity will be needed occasionally, one could retain the present arrangement and operate the suitable pump.

■ EXAMPLE 22.6

Water is delivered by gravity from a storage tank (liquid level elevation=62 FT), through a 6″ line (1,000 equivalent feet long) to another storage tank (liquid level elevation=0 FT). The capacity transferred is approximately 1,000 GPM.

It is desired to increase the capacity from 1,000 to 1,400 GPM. For the purpose, a new pump will be installed at level 0 FT (drawing water from a pond), and its new 6″ discharge line will be connected to the existing 6″ line, 400 equivalent FT from the receiving tank.

Size the pump (Q, H, BHP, motor HP) for a pump efficiency of 80% (Refer to fig. 22.17). Plot the system-head curve and the pump curve.

The hydraulic gradient required at point C for 1,400 GPM to flow through the 400-FT long 6″ stretch is 47.2 FT. (friction loss for 1,400 GPM in a 6″ line for water=118 FT/1,000 FT). The differential between

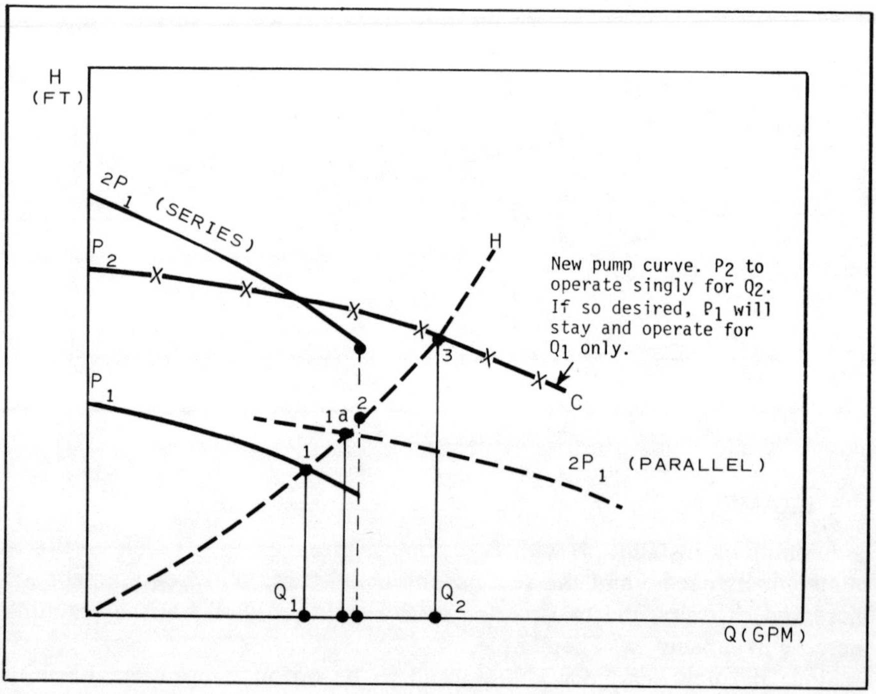

FIG. 22.16

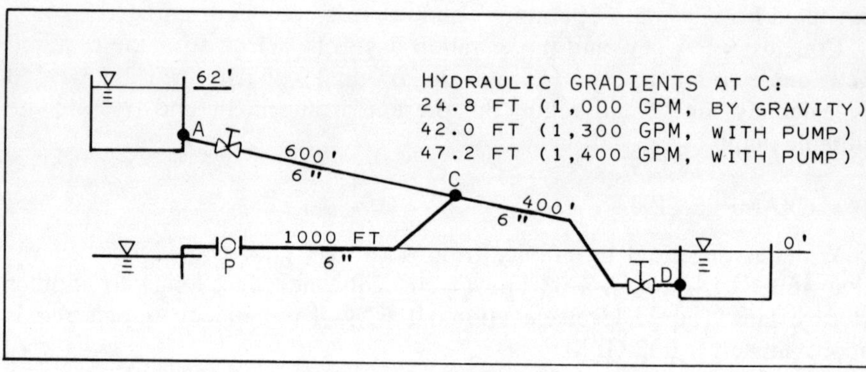

FIG. 22.17

the suction tank level and point C is now 62−47.2=14.8 FT. This will be the lost gravity head from the tank to point C, equivalent to the friction loss.

For the 600-FT stretch, this would correspond to a gravity flowrate of 620 GPM which leaves for the pump to supply 1,400−620=780 GPM. The friction loss for 780 GPM in the new 6″ pump line (1,000 FT long) will be 33 FT. The total pump head then equals friction loss PC+ hydraulic gradient at C minus pump elevation=33+47.2−0=80.2 FT. BHP=19.74, and motor HP=25.

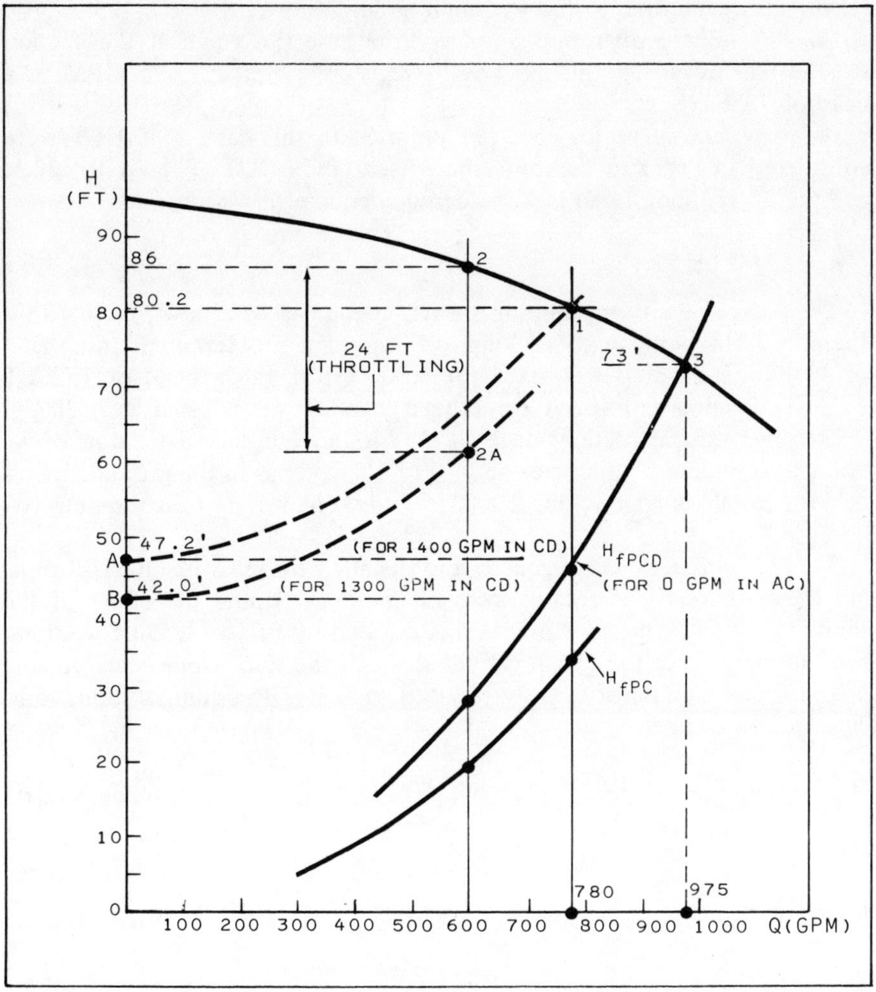

FIG. 22.18

Comments: (Figure 22.18)

Point 1=Pump operating point for outlined conditions: 780 GPM at 80.2 FT (47.2 FT hydraulic gradient, and 33 FT friction loss in PC).

A-1=System head, valid for this hydraulic gradient only.

Point 2=Pump operating point for throttled valve at the pump: 600 GPM at 86 FT (42 FT hydraulic gradient, 20 FT friction loss in PC, and loss due to throttling equivalent to 24 FT).

B-2a = System head valid for a hydraulic gradient of 42 FT only. (Determined by trial and corresponding to a flowrate of 700 GPM from A to C, and 1,300 GPM from C to D).
Note that the system head for an arrangement of this type is not constant. It will shift according to the hydraulic gradient at C.

Point 3 = Pump operating point for the case the valve at the suction storage tank is closed, and no flow is registered from A: 975 GPM at a head of 73 FT.

O-3 = System curve for case per point 3. In this case, 1,400 FT were considered in order to compute the losses (PC+CD). The BHP = 22.5, and the 25-HP motor can be considered adequate.

■ EXAMPLE 22.7

The design of a new pumping system encompassed a new pump with a spare, a 2,500 equivalent FT long 10″ line, and a differential static head of 40 FT. The pump was sized for 3,000 GPM and a head of 152 FT (112 FT friction and 40 FT static head), and it was selected accordingly.

During the construction stage, it was found out that a portion of an old, existing pipe, 6″ diameter and 400 FT long, was in the ground. It was decided to take advantage of it and use it by connecting to it the new 10″ line.

The line was cleaned, recoated and installed in line with the 10″ pipe. No computations whatsoever were made to determine the effect of the smaller portion of the pipeline. It was felt that 400 FT of 6″ line wouldn't substantially affect the pump output through the line. Upon start-up, the operator complained of sharply reduced capacity. The computations indicate the following:

$$H_{f10''} = f\frac{2{,}500}{10^5} \times \frac{Q^2}{32.18} = Q^2/80{,}450 \qquad \text{(for f=0.016)}$$

$$H_{f6''} = f\frac{400}{6^5} \times \frac{Q^2}{32.18} = Q^2/25{,}000 \qquad \text{(for f=0.025-old pipe)}$$

$$H_{f10''} = f\frac{2{,}100}{10^5} \times \frac{Q^2}{32.18} = Q^2/95{,}770 \qquad \text{(for f=0.016)}$$

H_{LINE} (10″ only) = $Q^2/80{,}450 + 40$ (2,500 FT long)

H_{LINE} (6″ & 10″) = $Q^2/25{,}000 + Q^2/95{,}770 + 40 = Q^2/19{,}820 + 40$

Points for the System-Head Curves:

Q (GPM)	1,000	2,000	3,000	4,000
$H_{10''}$ (FT)	52	90	152	239
$H_{6''+10''}$ (FT)	90	242	494	847

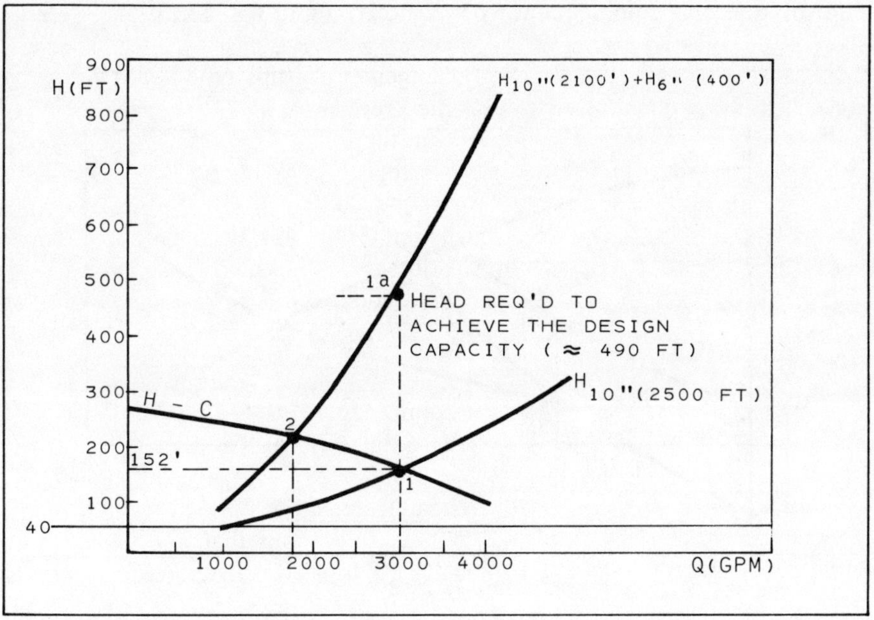

FIG. 22.19

Conclusion: Per fig. 22.19, with H−C curve selected as indicated, point 1 would have been the operating point, had the length of the 10″ line been left at 2,500 FT, as designed. The effect of the inclusion of the 400 FT of existing 6″ line and shortening the 10″ line used from 2,500 to 2,100 FT is indicated by point 2.

Instead of providing 3,000 GPM, the pump will provide only approximately 1,850 GPM, a mere 62% of the design flowrate. Also, the pump is operating much less efficiently.

It didn't pay to take advantage of the existing 6″ line without an investigation. With reference to chapter 4.3 and formula 4-06 ($D_2=D_1/n^{0.4}$): $n=(D_1/D_2)^{2.5}=(10/6)^{2.5}=3.58$. One should have used at least (3) 6″ lines in order to achieve a loss equivalent to the one in the 10″ line. It is less expensive to abandon the 6″ line.

■ EXAMPLE 22:8

An existing water transfer system consists of a storage tank connected to a receiving tank by a pipe. Flow is by gravity. The capacity resulting from the head differential is found subsequently to be inadequate. The installation of a pump is required based on (1) increasing the capacity to the existing receiving tank, and (2) considering simultaneous or individual future transfer to an additional storage tank situated at a different elevation. The solution and comments are per fig. 22.20 and fig. 22.21.

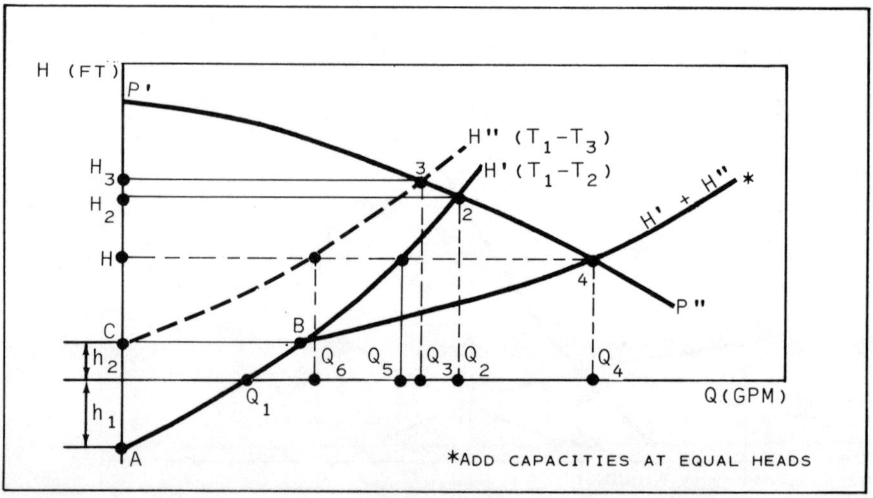

FIG. 22.20

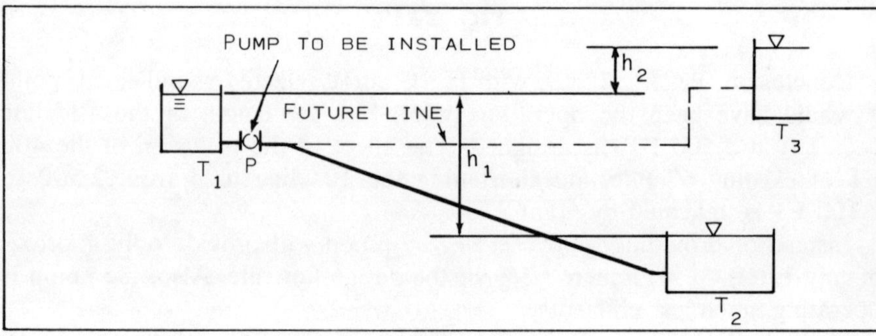

FIG. 22.21

Q_1=capacity by gravity (original).

Q_2=increased capacity to tank T_2 with new pump (individual operation, head=H_2).

Q_3=capacity to the future tank T_3 with future line connecting T_1 and T_3 (individual operation, head=H_3).

Q_4=capacity to both tanks (simultaneous operation in the future) with Q_5 to tank T_2 and Q_6 to tank T_3, (head=H); and ($Q_5 + Q_6 = Q_4$).

The selection of the pump must provide stable operating points for all cases, present and future. For the purpose, the layout of the future expansion must be known, so the computations would permit prediction of all operating points.

P′ − P″ = selected pump curve.
2 = operating point to tank 2 only.
3 = operating point to tank 3 only (future).
4 = operating point to both tanks simultaneously.
A − B − 2 = existing system-head curve (from tank 1 to tank 2).
C − 3 = system-head curve of future connection to tank 3.
A − B − 4 = combined system curve for simultaneous operation to both tanks. Note that no flow will be registered to tank 3 before the head for $T_1 - T_2$ exceeds $(h_1 + h_2)$, point B.

■ EXAMPLE 22:9

An existing pump takes suction from a tank and supplies product through a pipeline to a storage facility in a refinery about 2 miles away. Considering that the operation is intermittent, a decision is made to connect a new pipeline to the existing discharge line and supply barges at a pier, half a mile away, with the same product, as a secondary operation.

The size of the new pipeline is identical to the existing one. It is felt that any capacity which could be delivered to the pier would be satisfactory.

The result was that, when operating to the pier, the pump cavitated.

Comments: Per fig. 22.22, one had to be more selective with respect to the size of the new pipeline in order to obtain a steeper system head. The best solution for the case is to install a control valve for minimum allowable pressure so cavitation can be avoided.

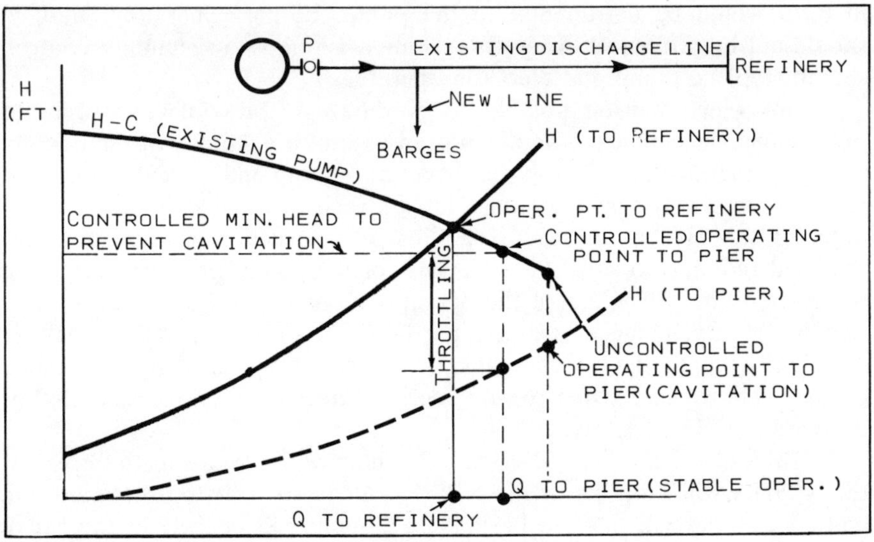

FIG. 22.22

23. Controls

One of the major advantages of a centrifugal pump is that its performance can be easily and simply controlled. Frequent mention has been made in foregoing examples of pump controls as a necessity or requirement to improve performance of the pump for the particular application.

▪ 1. CONTROLS TO PROTECT THE PUMP.

OVERHEATING: This will happen with the pump operating near or at shut-off pressure, as the result of partially or completely closed valve in the discharge line, any obstruction in it, or increase in total head for any reason. A by-pass with a relief valve to open at a preset pressure will permit product to recirculate, and prevent its temperature to increase which, in turn, would be detrimental to the pump. By-pass operation, though, should not be prolonged. Therefore a timer, acting as a secondary control, should stop the pump and alert the operator.

Pumps which transfer hot liquids may have to be cooled, in order to protect their mechanical components. In case of breakdown of the cooling system, a thermostat will stop the driver and pump and alert the operator. Larger pumps may sustain damage in case their bearings overheat.

SUCTION HEAD LOSS: If, for a reason, the suction head drops below the head required to satisfy the NPSHR of the pump, pressure sensors in the suction line will interrupt the operation. The reasons for a loss in suction head vary: clogged strainer, partially closed suction valve, lack of sufficient head in the suction tank or vessel, or higher than anticipated product temperature which would increase the vapor pressure, and thus decrease the NPSHA, etc.

OPERATING IN THE BREAK: A pump will operate in the break if the NPSHA will drop below the NPSHR. Controls will stop the operation and corrections will have to be made. A pump will operate in the break also if the system-head curve shifts during operation and doesn't anymore cross the pump curve within its stable range. This can happen in case, for example, there is a major break in the discharge line, and pressure drops suddenly. Minimum pressure-sustaining control valves will hold, for cases of this type, the operating point on the pump curve and prevent cavitation.

OVERSPEED: Pumps driven by turbines or engines can have their speeds increased, if the power of the driver becomes excessive and remains uncontrolled. The pump can be damaged. Overspeed governors control the speed of this type of drivers. Electric motors do not require speed controls, but they do require temperature sensors to protect them from overheating and thus breaking down.

REVERSE FLOW AND SURGES: Each pump must have a suitable check valve to protect the pump from reverse flow and surge pressures which may damage the pump. Reverse flow may also damage the driver. Excessive pressures may build up.

▪ 2. CONTROLS TO PROTECT THE PROCESS.

OVERFILLING: To prevent overfilling of a vessel, float controls will stop the pump, in case the operator will fail to do so.

MAINTAINING A PRESET PRESSURE, INDEPENDENT OF CAPACITY: This is achieved by means of a regulating valve, and its effect is shown on fig. 23.01. The head will remain constant, independent of the variable system head curve.

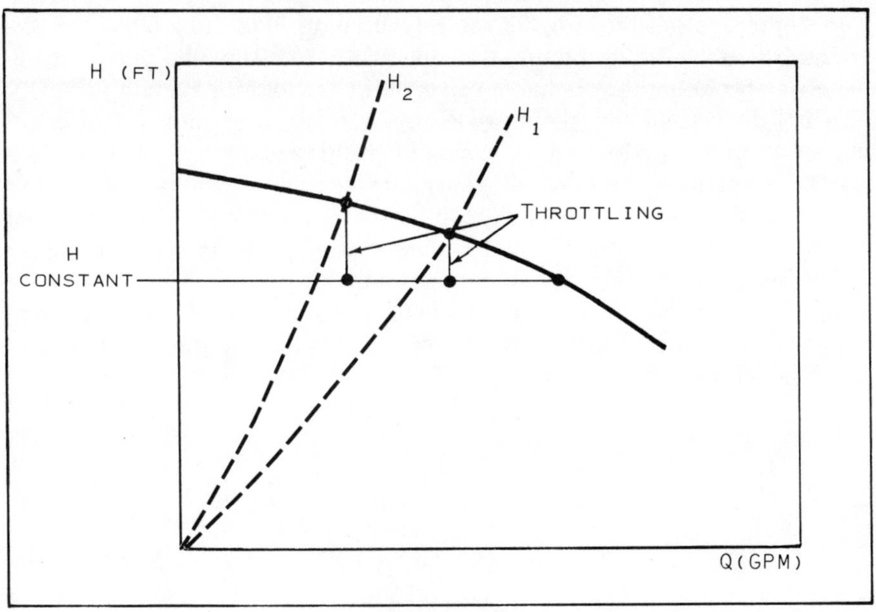

FIG. 23.01
Constant head, independent of capacity.

REQUIREMENT FOR INCREASE OR DECREASE OF CAPACITY: This may be necessary in order to maintain a specific process temperature in one or more elements in the system cooling water. Temperature sensors transmit a command to a control valve regulating the capacity. The operating point on the H−C curve will shift from higher to lower capacities and vice versa.

INTERRUPTION OF PRODUCT SUPPLY: For process reasons, the supply of product may have to be suspended. This is done by either stopping the pump, or by means of by-passing the product back to the suction source, or both, depending on the frequency of interruption required. A pump shouldn't be stopped and started too frequently.

▪ 3. EXCLUSION OR INCLUSION OF ONE OR MORE PUMPS.

A system in which multiple pumps are arranged to operate in parallel or in series, may, for various reasons, require, at a time or another, a substantial increase or decrease in capacity or pressure. This is done by timed, or programmed controls, designed to stop, or start one or more pumps in the pumping system.

▪ 4. CONTROLLING PROPER STARTING AND STOPPING OF A PUMP.

In most cases, the proper procedure to start a centrifugal pump of the volute category is to: (1) open suction valve(s), (2) start pump (with discharge valve closed), and (3) open discharge valve.
The stopping procedure for the same type pump is to: (1) close the discharge valve, while the pump is in operation, (2) stop the pump immediately after closing the discharge valve, and (3) close the suction valve(s).

If not done manually, these procedures can be programmed, and opening, or closing is performed by means of pushing a single button. The sequence is automatic (sequential starting or stopping). Note that pumps of the mixed-flow and propeller type, due to the very high shut-off power requirement, should preferably be started with partially open discharge valves, in order to avoid unstable operation.

Controlling centrifugal pumps is considered a specialty by itself, encompassing very modern and versatile control equipment, the discussion of which is beyond the scope of this book.

24.

Concluding Hints for the Application Engineer

1. There is only one way to approach pumping system design: thorough.

2. There is never an "easy" problem. Most of the troubles we have with pumping systems come from belittling "easy" problems. Design problems should be categorized only in relation to the time it will take to resolve them, and not according to degree of difficulty.

3. Do not get influenced by directives like "there is nothing much to it", or "the specification form has to be filled out and sent out as soon as possible." Remember that, if you hurry, you will make mistakes, and you will get no credit for your fast performance. Your good performance, though, will be taken for granted.

4. Upon attacking a pumping problem:

- Make a sketch covering the whole system.
- Enter all available physical dimensions and equipment data.
- Obtain all required product characteristics.
- Understand the rationale behind the requirements.
- In case a solution is already suggested to you, take the time (your own, if necessary) and analyze the system and the requirements. Come up with alternative solutions, evaluate them, and speak up, if you feel that you have a more economical and efficient system, or solution, to offer.
- Request as-built drawings, in case you will have to work with an existing system.
- Take a field trip, and talk to the operating personnel.

5. Try not to work out solutions from nomograms, charts; avoid shortcuts.

6. Equip yourself with manufacturers' catalogs on centrifugal pumps, valves, strainers, and piping fittings. For a specific problem, consult a catalog for a "K" factor, or equivalent length value rather than a handbook. Approach the manufacturer for specific questions.

7. Select better textbooks on centrifugal pumps and study them at a pace, allowing you to absorb the theory better.

8. Study carefully manufacturers' catalogs on various kinds of centrifugal pumps. They will supply you with a tremendous amount of information on pump types, sizes, construction details, and installation hints. Manufacturers' representatives are very helpful, too, when approached. They will be glad to discuss any specific problem with you.

9. Try to work out your pumping problems analytically, as in the preceding chapters of this book. Be careful with tabulated friction losses for different pipe sizes. You may easily be mislead.

10. Always convert units into feet of liquid, or absolute pressure. Try not to get entangled with inches of mercury or water, lift, vacuum, negative gauge pressures, etc.

11. Write down your calculations in order, spelling out the complete formulae in detail with the units for each parameter. List the numerical values in a separate column, or on the sketch. Describe the procedure concisely. The insertion of the numerical values in the formulae should be the last step, and not the first one.

12. Spend sufficient time analyzing the alternatives you come up with, the pump curves that would apply, and watch for pitfalls mentioned in previous chapters.

14. Read articles about centrifugal pumps in technical magazines; collect and classify them. Don't accept blindly that all statements made are correct, simply because the article has been printed. Use your own judgement.

15. Discuss pumping problems with other application engineers. Learn more about their specific problems and experience. Exchange opinions.

16. When you complete your work on a pump problem, read and check everything over very carefully. The number of corrections and changes you will make will amaze you.

List of Formulae and Relationships

Number

1-01 PSIA = Atmospheric Pressure + PSIG

1-02 1 Psi = 2.31 feet of water at 68° F.

1-03 0.433 PSI = 1.00 FT of water at 68° F.
(1 Foot of liquid = sp. gr. of liquid/2.31 psi)

1-04 1 PSI (any liquid) = 2.31/sp.gr. of liquid, in feet

1-05 1 inch of water = 0.036 PSI

1-06 1 inch of mercury = 0.049 PSI

1-07 Vacuum = 0.00 PSIA

1-08 1 CU FT = 7.481 Gallons

1-09 1 CU FT = 0.178 Barrel

1-10 1 Gallon = 0.1337 CU FT

1-11 1 Barrel = 5.615 CU FT

1-12 1 Barrel = 42 Gallons

1-13 1 Gallon = 0.0238 Barrel

1-14 Density of water at 68° F = 62.33 lbs/cu ft

1-15 Specific Gravity of Water at 68° F. = 1.00

1-16 1 Horsepower (HP) = 550 ft-lbs/sec = 33,000 ft-lbs/min

1-17 Torque = 5,250 × HP/ RPM

1-18 Centistoke = Centipoise/sp.gr.

1-19 V (FPS) = 1.273 Q/D^2, Q in CFS, and D in feet

1-20 $V\ (FPS) = Q/(352.6 \times D^2)$, Q in GPM, D in FT

1-21 $V\ (FPS) = Q/(2.448 \times D^2)$, Q in GPM, D in inches

1-22 $Q\ (GPM) = V \times 2.448 \times D^2$, V in FPS, D in inches

1-23 V_h = Velocity head = $V^2/2g$

1-24 Reynold's Number (RE) = $3{,}162 \times Q/(D \times CS)$, Q in GPM, D in inches

2-01 Friction Loss = Energy Loss = $K \times V^2/2g$, in feet

2-02 $K = f \times L/D$, dimensionless

2-03 $H_f = \frac{f \times L}{D} \times V^2/2g$, in FT; (L, D in feet, V in FPS)

2-04 $H_f = \frac{f \times L}{D^5} \times Q^2/8{,}006{,}658$ (in FT); (L, D in FT, Q in GPM)

2-05 $H_f = f \times L \times Q^2/(D^5 \times 32.18)$, in FT; (L in FT, D in inches, Q in GPM.)

3-01 $P_1 + V_1^2/2g + Z_1 + E_p - H_f = P_2 + V_1^2/2g + Z_2$, (all parameters in feet)

3-02 $f = 0.184/\ RE^{0.2}$ (for RE smaller than 1,000,000)

4-01 $H_f = (fQ^2/32.18) \times (L_A/D_A{}^5 + L_B/D_B{}^5)$

4-02 $H_f = (fQ^2/32.18)\ (L_1/D_1{}^5 + L_2/D_2{}^5 + \text{- - -} + Ln/Dn^5)$

4-03 $H_{f_1}/H_{f_2} = (f_1/f_2) \times (L_1/L_2) \times (D_2/D_1)^5$—for equal Q's

4-04 $H_{f_1}/H_{f_2} = (L_1/L_2) \times (D_2/D_I)^5$—for equal f's and Q's

4-05 $Le/L = f/fe \times (De/D)^5$ —for equal Q's

4-06 $D_2 = D_1/n^{0.4}$ (D_1 = larger pipe; D_2 = smaller pipe). —for equal Q's

5-01 $H_f = C \times Q^2$

5-02 $C = (fLe/D^5) \times 1/32.18$

5-03 $H_{f_1}/H_{f_2} = (Q_1/Q_2)^2$

7-01 $E_p = (P_2 - P_1) + (Z_2 - Z_1) + (Q^2/32.18)(f_1 Le/D^5 + f_2 le/d^5)$

8-01 $BHP = \frac{Q(GPM) \times 0.1337\ CF/GAL \times 62.33\ LBS/CF \times sp.gr. \times Total\ Head}{33{,}000 \times PUMP\ EFFICIENCY\ (\%,\ fraction)}$

8-02 $BHP = \frac{Q \times H \times Sp.gr.}{3{,}960 \times eff}$

8-03 $NPSHA = P_1 + Z_1 - fLsQ^2/(ds^5 \times 32.18) - Vp$; (s-suction line)

9-01 $D = 1{,}840 \times H^{0.5}/RPM$

11-01 $Q_2/Q_1 = D_2/D_1$

11-02 $H_2/H_1 = (D_2/D_1)^2$

11-03 $BHP_2/BHP_1 = (D_2/D_1)^3$

11-04 $Q_2/Q_1 = N_2/N_1$

11-05 $H_2/H_1 = (N_2/N_1)^2$

11-06 $BHP_2/BHP_1 = (N_2/N_1)^3$

11-07 $NPSHR_2/NPSHR_1 = (N_2/N_1)^m$; $m = 1.3\text{-}2.00$

11-08 $BHP = C_{QM} \times Q \times C_{HM} \times H \times sp.gr./(3{,}960) \times C_E \times Eff.)$

12-01 $NS = RPM \times Q^{0.5}/H^{0.75}$

12-02 $Q_1/(RPM_1 \times D_1^3) = Q_2/(RPM_2 \times D_2^3)$

20-01 $(P\text{-}L)\frac{i}{(1+i)^n - 1} + Pi + OC$

20-02 $Pi/(1+i)^n - 1) + Pi + OC$

20-03 $Pi + OC$

Recommended Literature:

Standards of the Hydraulic Institute, New York

Karrasik, Carter, Wright
Pump Questions and Answers
McGraw Hill Book Co., 1949

Karrasik, Carter
Centrifugal Pumps
Selection, Operation and Maintenance
McGraw Hill Book Co., 1960

Hicks, Edwards
Pump Application Engineering
McGraw Hill Book Co., 1971

King Brater
Handbook of Hydraulics
McGraw Hill Book Co., 1963

Baumeister and Marks
Standard Handbook for Mechanical Engineers
McGraw Hill Book Co., 1967

Crocker and King
Piping Handbook
McGraw Hill Book Co., 1967

E. E. Ludwig
Applied Process Design for Chemical
and Petrochemical Plants, Volume I
Gulf Publishing Company, 1964

Westaway, Loomis
Cameron Hydraulic Data
Ingersoll Rand Company, 1977

Karassik, Krutzsch, Frazer, Messina
Pump Handbook
McGraw Hill Book Co., 1976

Holland, Chapman
Pumping of Liquids
Reinhold Publishing Corporation, 1968

Index